MATHEMATICS
IN ACTION

MATHEMATICS IN ACTION

Applications in Aerodynamics,
Statistics, Weather Prediction
and Other Sciences

By

O. G. SUTTON, KT.

C.B.E., D.Sc., F.R.S.

WITH DRAWINGS AND DIAGRAMS BY

A. J. LAURIE

DOVER PUBLICATIONS, INC.
NEW YORK

This Dover edition, first published in 1984, is an unabridged and unaltered republication of the 2nd, 1957, edition of the work first published by G. Bell & Sons, Ltd. in 1954
under the title *Mathematics in Action*. It is reprinted by special
arrangement with Bell & Hyman Ltd., 37/39 Queen Eliz-
abeth Street, London SE1 2QB England.

Manufactured in the United States of America
Dover Publications, Inc., 31 East 2nd Street, Mineola, N.Y.
11501.

Library of Congress Cataloging in Publication Data
Sutton, O. G. (Oliver Graham)
 Mathematics in action.

 Reprint. Originally published: London : G. Bell and Sons,
1954.
 Includes index.
 1. Mathematics—Popular works. I. Title.
QA93.S87   1984   510   84-8022
ISBN 0-486-24759-7

PREFACE

THIS BOOK, written primarily for the layman, will prove, I hope, of interest also to students, especially those in the upper forms of schools or in their first years in the university. It is a view of the part played by mathematics in applied science, as seen by a mathematical physicist. The topics which are discussed in detail are all taken from subjects in which I have worked, professionally, from time to time. This explains what may appear otherwise to be a random selection of material from a very large field.

It is a pleasure to acknowledge my indebtedness to Mr. S. L. Dennis who suggested the theme, to my colleagues Professor G. D. West, Dr. C. G. Lambe and Mr. R. B. Scott, who read and criticized some of the chapters in manuscript, and to Mr. A. J. Laurie, who provided the illustrations from the roughest of sketches. My thanks also go to my wife for her help in checking and proof reading, and to Miss E. J. Smith for preparing the typescript.

O. G. SUTTON

Shrivenham, Berks
July, 1953

PREFACE TO THE SECOND EDITION

IN THIS edition, Chapter 7 has been enlarged to include a fairly detailed account of numerical forecasting, which has now reached the operational stage. I have also added a short Appendix on the calculation of probabilities in coin-tossing trials.

I am indebted to correspondents and others who kindly drew my attention to some errors and ambiguities in the original text. These have been rectified.

O. G. S.

June, 1956

CONTENTS

INTRODUCTION

MISS SUSAN: *What is algebra exactly; is it those three-cornered things?*
PHOEBE: *It is x minus y equals z plus y and things like that. And all the time you are saying they are equal, you feel in your heart, why should they be?*

J. M. BARRIE, Quality Street

The mathematician, carried along on his flood of symbols, dealing apparently with purely formal truths, may still reach results of endless importance for our description of the physical universe.

KARL PEARSON

THE PROFESSION OF mathematics is much like any other, with its ups and downs, its moments of excitement and years of drudgery, and if the labours of mathematicians differ from those of their fellow scientists, it is only in the abstract and seemingly remote nature of the subject. Mathematics, like music and poetry, is a creation of the mind; its reality is not that of the physical world, and the primary task of the mathematician, like that of any other artist, is to extend man's mental horizon by representation and interpretation. Yet this difficult and austere branch of learning is the keystone of the arch which science has built to span the gulf between the world of sensory impressions and the ultimate reality. A defence of mathematics, if such be needed, may well start from the premise that 'the proper study of mankind is man', for the triumph of mathematics in physical science is in many ways the story of man's conquest of nature.

This book is concerned solely with mathematics in physical science or, in more homely terms, with 'applied', as distinct from 'pure', mathematics. In many ways this division is artificial. It is true that in the minds of most mathematicians there is a real distinction between the two kinds of mathematics, but the difference is never absolute. Thus hydrodynamics, the study of the motion of fluids, is in some ways so highly abstract that many engineers regard it as a part of pure mathematics. One of the most 'pure' of all branches

1

of mathematics is geometry, which sprang from the prosaic problem of fixing the boundaries of fields in ancient Egypt. The geometry which has been derived from such simple considerations is called Euclidean; it is that which we study in school and afterwards use to lay a carpet, plan a garden or survey a site for a new town. To the mathematician it is only one of many geometries (and a rather dull one at that), all of which deal with the interrelations of complexes of abstract conceptions called, mainly for convenience, 'points', 'lines', 'figures' and 'surfaces'. These geometries are in no way necessarily related to the everyday world of fields, plans and boundaries, and geometers, as a class, are unconcerned about such lack oₓ touch with 'reality'. Occasionally it happens that someone, perhaps Einstein, uses one of these geometries to assist him in forming a picture of the world which lies about us and beyond the stars. If such an attempt is successful, the abstract geometry of the pure mathematician acquires a new importance and may be regarded, justly, as a part of applied mathematics, but this does not mean that a particular geometry alone is true and others are false. If the pure mathematicians know their job, all their geometries are equally true, but some are thought to correspond more closely to the universe of space and time than others. In pure mathematics this correspondence is irrelevant, but in applied mathematics it matters more than anything else.

Mathematics is a study of patterns of ideas, executed by means of a highly developed specialized technique, which is believed to be infallible. It is concerned only with abstractions. Since the days of ancient Greece we have come to rely more and more upon such study as the only sure way of gaining mastery over nature, and the process is nearly always the same in its essential features. The pure mathematician follows a certain line of thought, inventing the necessary techniques as he proceeds, without paying much heed to possible applications in the physical world. From time to time it happens that a particular development in pure mathematics is found to bear a peculiar relation (often not very obvious) to a problem in physics. Or it may happen that a problem in physics can be recast in a form suggested by the theorems of the pure mathematician. The task of the applied mathematician is exactly that of using the tools provided by pure mathematics to clarify and extend the observations of the physicist.

The purpose of this book is to show this process in action. Strictly, mathematics cannot reveal anything about the world of concrete objects, because it deals exclusively with mental concepts, but it is equally true that in physical science, as Darwin says, 'Every *new* body of discovery is mathematical in form because there is no other guidance we can have'. Sooner or later a phenomenon which has appeared, perhaps unexpectedly, in the laboratory must be measured and reduced to numbers if it is to find a permanent place in the body of scientific knowledge. Until this is done, it is like some wild animal which has emerged from the depths of the unplumbed universe around us.

The difficulties commonly experienced in explaining mathematics to a non-mathematical audience are, in many ways, inherent in the subject itself. Mathematics is economy in thought carried to extremes; it is devoid of all the emotions and associations which affect most other acts of thinking, and any explanation in general terms must inevitably introduce extraneous elements which tend to falsify or cloud the true picture. The lay reader must be content with the general view, leaving the details to the experts ; but since it is mainly the details which demand specialized training, a broad understanding of the general method and the results obtained by mathematics is not difficult to achieve.

A technique succeeds in mathematical physics, not by a clever trick, or a happy accident, but because it expresses some aspect of a physical truth. This is the real answer to Phoebe's question, and part of the purpose of this book is to disclose, as far as are known, the causes which underlie the successful treatment of a physical problem by mathematics. Another recurrent question, implicit in the quotation from Karl Pearson, lies far deeper—how can the manipulation of symbols which *we* have invented, according to rules which *we* alone make (and sometimes break), reveal that which lies beyond our senses? In other words, how can we discover anything by mathematics? Whether or not the reader holds, with Hardy, that 'mathematical reality lies outside us ...

the theorems which we prove, and which we describe grandilo-
quently as our "creations", are simply the notes of our observations',* the question is one which is unlikely to receive a
satisfactory answer in this generation, or perhaps for many
years to come. We can, at the most, indicate a few possible
clues to the answer by a general view, such as that given in
the pages which follow.

* G. H. Hardy, *A Mathematician's Apology*, p. 63 (Cambridge, 1940). Hardy, of
course, is thinking mainly of discoveries in pure mathematics.

CHAPTER

1

THE MATHEMATICIAN AND HIS TASK

'Imaginary' universes are so much more beautiful than this stupidly constructed 'real' one; and most of the finest products of an applied mathematician's fancy must be rejected, as soon as they have been created, for the brutal but sufficient reason that they do not fit the facts.

G. H. HARDY

The Theories of Physics

ANYONE WHO GLANCES at current literature on mathematics and physics is bound to be struck by the frequency with which the word 'theory' appears. In pure mathematics there are the theory of functions, the theory of numbers, the theory of differential equations and many more; applied mathematics or mathematical physics abounds in theories of the expanding universe, of turbulent motion, of atomic structure, to quote only a few. It is evident that the word is being used in two different senses. In pure mathematics, 'theory' invariably denotes an exposition of the fundamental principles of a subject, and the main facts, called *theorems*, are to be found in the same or completely equivalent forms in all accounts. Only the logical sequence is liable to vary from one author to another. In physical science, on the other hand, the word more often denotes a sphere of speculative thought, and the assertions of one scientist may be received with scepticism, or even vehemently denied, by another. This divergence of opinion seems to reach an extreme level in the biological sciences, and an outsider sometimes may be inclined to the belief that the protagonists spend more time in belabouring each other than in seeking the truth.

This book is primarily concerned with theories of the physical universe, using the word in the sense of speculative thought. A physical theory is a rational explanation of how the world of

5

material objects is constructed, acquires its characteristic properties and continues to exist. The word *how* is essential; a physicist does not attempt to explain *why*, leaving that aspect to the philosopher and the theologian.

Physical theories of one sort or another are as old as mankind. They arise from man's inherent belief that every event is uniquely related to some immediate antecedent event, called a *cause*, and to some subsequent event, called a *consequence*. A man who is hit on the back by a stone is unlikely

FIG. 1.   Two Theories of the Eclipse of the Sun

to believe that the blow occurred, as it were, on its own; he is certain to look for someone who threw the stone, and ultimately he associates the pain he feels with the act of throwing the stone. In arranging the events mentally in this order, and in disregarding certain simultaneous events, such as the appearance of a cloud in the sky, as irrelevant, the victim is forming a simple and plausible theory to account for his physical sensations. In primitive peoples this process is carried to extremes and all events are related to an animistic universe. In some parts of the world it is still believed that a solar eclipse is caused by a dragon trying to swallow the sun, and attempts are made to avert the threatened catastrophe by exploding crackers and beating gongs to drive away the dragon. This is a physical theory of a very crude type, but it is at least an attempt to associate a somewhat rare astronomical event with a more common experience (namely, eating). The theory is perfectly satisfactory provided one is prepared to believe in dragons, but is rejected by more advanced thinkers for a variety of reasons, among the more important of which are: (a) no one with normal eyesight and in a state of sobriety has ever seen a celestial dragon, (b) other observations make it difficult to believe that any living creature could swallow the sun and (c) there is a simpler explanation, based on certain data relating to the motion of spheres in elliptic orbits which not only bring such events within the range of human experience, but also allow the recurrence of eclipses to be predicted with certainty. The modern theory, it should be noted, relies mainly on (a) bringing in only observable entities, (b) consistency between the proffered explanation and other independent physical measurements, and (c) the fact that the mathematical consequences of the theory are found to agree with repetitions of the same phenomenon in all parts of the world. In addition, the astronomical theory has the important merit, to western eyes, of being the *simplest* yet advanced.

Physical theories, to be acceptable to the twentieth-century scientist, must have all these features in varying degrees. Primitive man met his intellectual difficulties by postulating a god or demon as the cause of everything he could not understand, from earthquakes and thunder-storms to plagues and failures of crops. The modern attitude calls for the unhesitating use of 'Occam's razor', *Entia non sunt multiplicanda sine necessitate* ('Entities are not to be multiplied unnecessarily'), but the process of expelling the fairies has been a long one. It is not so very many years ago that Benjamin Franklin, in an experiment that few would care to repeat to-day, drew electricity from a thunder-cloud by means of a kite, and so showed that Jove's thunder-bolts and the tiny spark from a condenser are one and the same thing. 'Physics is experience, arranged in economical order', wrote Mach, and a physical theory is an indirect description of a natural process, in which new facts are exchanged for old. In all physical theories, however novel they may seem, the basic process is the replacement of new, strange facts by arrangements of simpler and more familiar facts, which in some ways can represent the later discoveries.

New Lamps for Old

To illustrate the ever-changing nature of physical theory, let us consider what happened in the theory of heat. One of the most obvious facts about heat is *conduction*; a poker placed in a fire ultimately becomes hot everywhere, and not only at the point. It is natural to suppose that 'heat' has moved from the fire into the metal rod, and from this it is equally natural to deduce that heat really is a kind of fluid ('caloric'), the presence of which in a substance makes it hot. We still use the conception of the 'flow of heat' in mathematical work, the direction of flow being from high to low temperature, just as water flows downhill, and one of the most valuable and significant methods of analysing the effects of heat, the so-called reversible cycle, was evolved by an engineer (Carnot) who firmly believed that heat was a material substance. But the conception cannot be maintained, and in 1799 Rumford showed that although a great deal of heat is needed to melt a large quantity of ice, no change in weight can be detected in the resulting water. He speculated that heat was no more than 'a vibratory motion of the constituent parts of heated bodies', and in 1847 Joule, by measuring the heat generated by a known amount of work in agitating water by paddles, showed experimentally that heat and mechanical work are interchangeable at a fixed rate. This gave the death-blow to the caloric theory, useful though it may have been, and at the same time opened the door for the mathematicians. Kelvin developed the idea of heat as a form of energy; Clausius and Clerk Maxwell began to formulate the kinetic theory of gases, in which everything is explained in terms of molecules resembling little billiard balls continually colliding with each other, and finally heat became simply an expression of the energy of motion of the molecules. This is the view held to-day, so that the mystic, imponderable fluid caloric has been replaced by something more easily grasped, but we still find it convenient to use some, at least, of the fluid theory ideas.

A theory in physics lives just as long as it can evade contact with some irrefutable but contrary fact of experiment. The collapse of a theory is essential for progress, for, as Jeffreys says, it is the exception which *improves* the rule. A physical theory is seldom utterly wrong in its first form, but it cannot give more than a partial view of the process and, as observation proceeds, sooner or later it must be replaced by a more general doctrine. Perhaps the most celebrated example of this process of growth is afforded by the development of relativistic mechanics from ordinary (Newtonian) mechanics. The laws of motion as propounded by Newton are still valid for most applications, because the differences between ordinary and relativistic mechanics are so small that in most cases they are far beyond the discrimination of even the finest measuring instruments. It is only when considering phenomena on an immense scale (as in astronomy) or at very high speeds (as in certain aspects of nuclear physics) that the differences become measurable. The importance of Einstein's theory lies more in its physical interpretations than its practical consequences, and without Newton, relativity could never have been imagined, let alone made into a workable scheme.

This apparent impermanence of theory has produced an impression in many able minds that science is fundamentally unstable, for ever casting off the old in favour of the new; but this is to misunderstand the scientific method. One may as well reject a child because it is growing up. It is true that science has no everlasting dogmas, no sacred writings and no reverence for its early Fathers to the extent that a text from, say, the *Principia* is an absolute reply to an argument; but, on the other hand, when an inescapable fact shatters a time-honoured theory, there is usually something of the old in the new doctrine which emerges from the debris. It is commonplace to talk of a revolution in scientific thought caused by some new discovery, but it would be more accurate to speak of the change as evolution. Scientific progress is like mounting a ladder: each step upward is followed by a brief pause while the body regains its balance, and we can no more disregard the steps which have gone before than we could cut away the lower part of the ladder.

The Necessity of Mathematics in Science

In the exploration of the physical universe by scientific methods there are certain requirements for which mathematics is essential. To establish a new result a considerable amount of evidence must be collected, purified and put into a form in which it can be easily scrutinized. The accomplishment of the first two is the main task of the experimentalist. He must first design apparatus in which the phenomenon or process which he seeks to investigate is freed as far as possible from unwanted effects. If, for example, a physicist seeks to measure the earth's magnetism, he must take precautions to see that his readings are not rendered invalid by the intrusion of the magnetic field of some body whose presence in his laboratory is purely fortuitous. Secondly, he must reject any spurious readings and re-examine any that are suspect, and he may have to repeat his experiments many times before he can be sure that he has obtained results of real value. Much of the time spent in laboratory exercises is designed to develop in the young scientist a set of 'conditioned reflexes' to protect him against errors of this kind. The first task of mathematics begins only when the experimenter is thoroughly satisfied with his results. This is to provide *a convenient and accurate method of summarizing experience*, unless of course he is testing a theory which already has been cast in mathematical form. This task of mathematics, although humble, is of fundamental importance, for without some condensation of this kind it would be impossible for the mind to grasp the complicated pattern revealed by the laboratory experiments, and science would be hopelessly bogged in an ever-increasing mass of seemingly unrelated facts.

Empirical Formulæ in Physical Science

A mathematical expression which summarizes results without relating them to known scientific laws is called an *empirical formula*. There are hundreds of such formulæ, many of which are in daily use in physics, chemistry and engineering. Some of these have been devised to fill gaps left by theoretical analysis; others are scientific curiosities, in that no one, so far, has been able to build them into a coherent theory.

Perhaps the best-known example of a pure empirical formula is Bode's Law in astronomy. The planets in the solar system move in nearly circular orbits whose mean distances from the sun do not follow any very obvious rule. Yet there is a rule, although it is not quite exact. Take the numbers

$$0, 3, 6, 12, 24, 48, 96, 192, 384$$

that is, a sequence in which every number, except the first, is exactly half the number which follows it. Add to each the number 4, divide by 10 and call the mean distance of the earth's orbit 1. The resulting numbers agree fairly closely with most of the relative mean distances of the planets, thus:

Planets . . .	Mercury	Venus	Earth	Mars	Asteroids
Actual relative distance	0·39	0·72	1·00	1·52	2·77
Bode's Law . .	0·40	0·70	1·00	1·60	2·80

Planets . . .	Jupiter	Saturn	Uranus	Neptune	Pluto
Actual relative distance	5·20	9·54	19·19	30·07	39·60
Bode's Law . .	5·20	10·00	19·60	38·8	fails,

The law breaks down for Neptune but, curiously, the predicted mean distance for Neptune is almost exactly that found for the planet Pluto—a remote, cold world, about as big as the Earth, first located in 1930. Bode's Law was discovered—or perhaps stumbled on—as long ago as 1778, and has remained an enigma ever since. To-day no one can say with certainty whether it is a mere accident of figures or an indication of a deep physical law as yet undiscovered.

A relation of this type resembles the strange rules which some people use to pick numbers at roulette or teams in a football pool, but in physical science the experimenter usually has some assistance from theory. The most important guiding principle is that of forming *dimensionless groups*, and this requires some further explanation.

It is an axiom of physical science that a genuine law of nature must be capable of being expressed in a form independent of the units in which the measurements are made. The reasonableness of this is almost self-evident. No one would give a second thought to Bode's Law if it held only when the distances were measured in miles, for miles have no significance outside the British Commonwealth and America. The units in which the planetary distances are expressed are immaterial for Bode's Law, for the rule relates to *ratios* of the distances, which must be the same in all units. All physical quantities, such as speed, density, momentum, are derived from certain *fundamental magnitudes* capable of being measured directly. These are chiefly mass, length and time. Most physical magnitudes cannot be measured directly; they are the *derived magnitudes* which appear in physical laws. A velocity, for example, is the length covered in unit time, and density is the mass of unit volume; we say therefore that the *dimensions* of velocity are length divided by time, and of density, mass divided by volume (that is, by the cube of a length). When we have chosen the *units* (centimetres or inches, pounds or grams) we can describe any fundamental or derived unit by a *pure number*. The number depends on the units and changes with them; thus the density of water is 1 gram per cubic centimetre or 62 pounds per cubic foot.

A mathematical identity is a relation between pure numbers, like 1, 2, 3, . . ., e, π, and any physical magnitudes which occur in such relations must be disposed in groups which themselves are pure numbers. This can be done by arranging the groups so that they are quotients of magnitudes having the same dimensions. Some of the more important of these dimensionless groups, which have the same numerical value in all systems of units, occur in fluid mechanics. The quantity which dominates supersonic flight is the Mach number, the ratio of the speed of the aircraft to that of sound-waves at the prescribed temperature, and it does not matter if the speeds are measured in miles per hour or centimetres per second. Another very important group is the Reynolds number, Vd/v, where V is a speed, d is a length and v is the kinematic viscosity, the physical magnitude which measures the resistance of the fluid to deformation. An experimenter investigating the flow of fluids in pipes will conclude that the magnitudes which matter most are the pressure (p), the density (ρ), the speed (V) and the viscosity (v) of the fluid, and the length (l) and diameter (d)

of the pipe. His task is to find some mathematical relation which summarizes his measurements. He begins by arranging his physical magnitudes in dimensionless groups:

$$\frac{Vd}{\nu}, \quad \frac{l}{d}, \quad \frac{p}{\rho V^2}$$

His next step probably will be to plot his results on graph paper, using pairs of groups as variables, and if the points are such that it is easy to draw a single smooth curve through them, he is assured that he has found something significant. A curve is equivalent to a mathematical formula, and is thus an empirical relation, but for the purpose of communicating his discoveries to his fellow workers he will probably decide that a mathematical expression is more convenient. The equation

$$\frac{p}{\rho V^2} = \frac{l}{d}\left\{a\left(\frac{\nu}{Vd}\right)^{0.35} + b\right\}$$

is one such expression, for flow in smooth pipes. Here a and b are pure numbers, depending on the smoothness of the pipe. Such an empirical formula sums up in one line of print the results of months, perhaps years, of work in the laboratory.

The best empirical formulæ are all of this type, concise, logical and simple. The fitting of a mathematical formula to a mass of figures is not significant in itself, because it can always be done in some form or other. The essential requirements are that the formula shall be simple and contain as few 'arbitrary constants' (that is, numbers which cannot be deduced from the physical magnitudes) as possible. A 'portmanteau' formula, containing three, four or more arbitrary constants, is unlikely to prove of great value for subsequent investigations.

The mathematics involved in constructing empirical formulæ is usually of the most humble kind, and for this reason such formulæ tend to be despised. This is a wrong view, for these formulæ, properly constructed, can point the way to deeper considerations. Sometimes such expressions indicate extraordinary insight; Balmer's formula for the lines in the spectrum of hydrogen was later triumphantly verified by Bohr's theoretical study of atomic structure, and there are many more examples to be found. When theory is brought in to help in the selection of the most significant variables and to indicate how they are grouped, the empirical formula can be, and usually is, an essential first step on the road towards clearer understanding of a physical process.

Mathematical Abstractions

The second use of mathematics in physical science is much more sophisticated, and brings into play the full range of mathematical techniques. This is *the solution, by mathematics, of problems in physics,* but we must first look a little more deeply into the exact meaning of this phrase. Mathematics, as we have noted, is concerned only with abstractions, and therefore has nothing to say about the spatio-temporal universe. Physical observations, however carefully made, can never 'prove' or 'disprove' a mathematical theory, because the two have nothing in common. Yet statements such as 'recent astronomical observations have shown that Einstein's theory is correct' are by no means uncommon in scientific literature.

The process by which mathematics is brought in to develop a physical theory is nearly always as follows. First, the physicist specifies a *real problem.* This may be, for example, the investigation of temperature inside a block of iron when the surface is heated to a certain temperature, or the distribution of electromagnetic waves around a radio transmitting aerial. To deal with such problems, we need, in the first place, certain established facts, such as the dimensions of the block of iron and the conductivity of the metal, or the geometrical shape of the aerial and its height above ground. We call this the *collection of data.* The next step usually demands considerable skill and experience. This is the setting of an *idealized problem* which must satisfy two essential requirements: first, it must bear a strong resemblance to the real problem, and second, it must be in a form to which existing mathematical techniques can be applied. The idealized problem is thus an abstraction from the real problem, and is the key to the whole process. The *solution of the ideal problem by mathematics* follows, after which there is a *comparison of the solution with measurements made in the real problem.* If the numbers appearing in the mathematical solution (say, the temperature in the iron or the field strength in the space around the aerial) agree with the numbers indicated by the thermometer or the electrical instruments, we are entitled to say, with some confidence, that the physical problem has been solved mathematically, or perhaps that a theory has been verified. The process is shown diagrammatically in Fig. 2.

To illustrate this process, consider a homely example. In the course of my duties I sometimes leave my college to visit certain research establishments. In other words, I travel by car from Shrivenham to, say, Malvern. The *real problem* which confronts me is that of finding how long the journey will take. This is quite insoluble by mathematics, but I can deal easily with a corresponding ideal problem. I start by *collecting data*; I find that the Ordnance Survey has measured the distance involved—about 60 miles—and recent observations on

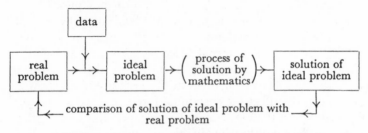

Fig. 2.   Schematic Representation of the Mathematical Solution of a
Problem in Physics

the state of my car support the belief that this somewhat aged vehicle is still capable of speeds up to 55 miles per hour. At this stage the real places Shrivenham and Malvern, the real road through Cheltenham and Tewkesbury and my real car vanish from the problem. Instead, I have an *ideal problem* of a point (representing my car) moving along a line 60 miles long (representing the road) at a constant speed of 30 miles per hour, the latter figure being one which experience has shown me to lead to good predictions of times spent in travel. I apply a known mathematical technique (division) to this problem, and obtain the answer 2 hours; my 'theory', if you like, is that the average speed on such a journey is 30 miles per hour, and this is verified by the measurement of time by my watch.

This example is, of itself, trivial, but the sequence described above occurs over and over again in mathematical physics. The essential feature is the creation of the abstract system to which mathematics can be applied. Non-mathematicians are often puzzled by the amount of attention given in mathematical texts to problems about weightless strings, frictionless fluids, perfectly rigid bodies and the like, none of which occurs in nature. It is true that many examples are simply exercises designed to develop facility in techniques, but apart from these, the study of imaginary systems is essential for a complete understanding of the real system. The task of the mathematical physicist is to make predictions. For this he needs an abstract system, an imaginary world which, it is hoped, resembles the real world closely in certain important features, but in designing his system the mathematician is prepared to exclude certain essential features of the real world as irrelevant, in order to bring other factors within the range of his techniques.

A good example of this process of idealization by exclusion is afforded by aerodynamics. Like all gases, air possesses a certain amount of internal friction, or viscosity, and the possibility of mechanical flight turns, in the end, on this particular property. Changes in the viscosity of air, however, do not greatly affect the sustaining force on a wing once flight has begun, and the mathematician seizes on this fact to effect a considerable simplification in his analysis. He postulates an imaginary atmosphere, totally devoid of friction, through which imaginary wings move at uniform speed. It is fairly easy to calculate the magnitude of the lift, or sustaining force in this system, and to make predictions how this force will vary with speed. Despite the removal from the picture of the one factor which makes flight possible, the predictions are found to agree well with measurements, and a notable advance in the science of flight was achieved in this way. On the other hand, the same hypothetical atmosphere proves to be useless for the calculation of the direct resistance, or drag, which the wing experiences in flight, since in an atmosphere devoid of viscosity the mathematical analysis leads to one result only: that drag is exactly zero for all bodies. Another model is needed in this instance, and the analysis is much more difficult.

Mathematical Techniques

At this point we must pause to consider an important distinction. The mathematician may construct imaginary worlds in order to explore certain features of the real world, but he

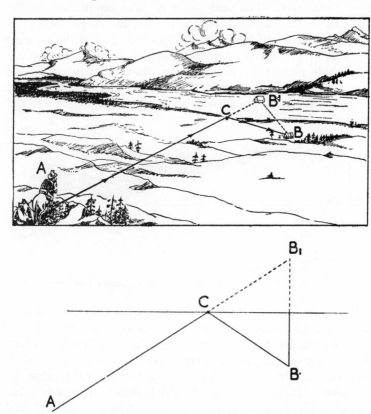

FIG. 3.   The Problem of the Traveller and the River

may also use his imagination to facilitate the solution of his problems.   In this he is like a draughtsman who puts in certain lines at the early stages, only to erase them in the finished drawing.   Let us consider how this is done in mathematics, by means of a very simple example in the first place.

A traveller, at a point A near a river, is on his way to his tent at B on the same side.   However, he must reach the river to water his horse before going to B, and his problem is to find the shortest route. If he knows a little geometry he can solve the problem very easily; he imagines that the bank of the river is like a mirror, in which B is reflected at B'. The shortest distance from A to B' is a straight line which intersects the river bank at C. His real route is then determined as the broken line ACB, and this road is shorter than any other.

This problem can be solved without introducing the imaginary tent at B', but the 'mirror' method is much the easiest. The whole picture is, of course, an abstraction in which the river is replaced by a straight line and the tent by a point, but the reflection at B' is a further mental artefact which does not appear in the solution.

Fig. 4.   The Problem of Smoke from a Factory Chimney

Now consider a much more difficult problem in which a very similar device is used. Smoke from a factory chimney is blown away by the wind as a long plume which at first keeps well above the ground. At a certain distance down wind (depending on the height of the chimney and meteorological conditions) the smoke reaches the ground, which forms an impenetrable barrier. The downward diffusion of the smoke is checked, and from this point onward the concentration (that is, the mass of smoke in unit volume of the atmosphere) is affected by the presence of the ground. The problem is to calculate the concentration at any point, given the rate of emission of smoke and certain meteorological data.

The problem can be solved in various ways, but the simplest method is to suppose that the ground is removed and that the atmosphere is continued into the space occupied by the earth. In this atmosphere there is placed a hypothetical source of smoke which is the mirror image of the real source in the line representing the ground, emitting a plume which is the mirror image of the real plume (Fig. 4). It is easy to see that in this system there is no *net* transport of smoke across the line which represented the surface of the earth, because for every smoke particle which crosses in one direction, a similar particle crosses in the other direction. By this device we have produced a system in which the line representing the ground has all the properties of an impervious boundary, and to solve the problem all we have to do is to combine the expressions for concentration from both sources. The result represents the distribution of smoke from a source at a given height above ground, and in the presentation of the result the 'reflected' source does not appear. It is simply a mathematician's device for getting the answer with the minimum of labour.

(This problem is of considerable interest in studies of atmospheric pollution and shows, in particular, that the concentration at any point varies inversely as the square of the height of the chimney. That is, if a certain concentration of smoke is found at a certain distance down wind of a 100-foot chimney, raising the height to 200 feet would reduce the concentration at this distance to one-quarter of what it was originally. This illustrates precisely the value of tall stacks in getting rid of unwanted fumes.)

Applied mathematics abounds in problem-solving devices of this kind. In hydrodynamics, and in problems of heat conduction, the mathematician makes frequent use of what are called 'sources' and 'sinks'—points at which fluid or heat is supposed to be emerging or disappearing. The original mathematical problem and the final solution usually contain no hint of such features, and in this sense sources, sinks, doublets and the like are simply the 'construction lines' of the mathematician in his task of building solutions.

Mathematical Worlds

Let us now return to the problem of idealizing a physical process, as distinct from a device to facilitate the discovery of solutions—that is, to investigations in which the properties of the system proposed affect significantly the conclusions reached.

Physicists are accustomed to assess the value of such idealizations by the closeness with which the properties of the model, as revealed by mathematical analysis, agree with features of the real world revealed by experiment and observation. (Pure mathematicians, on the other hand, are more likely to be swayed by æsthetic considerations).* The great respect which is accorded to mathematics in the modern world rests largely on the fact that in many instances this correspondence is extremely close. To take a simple example, the calculation of the height of a flagstaff from observations made at some distant points with a theodolite and a measuring tape agrees so well with the direct measurement of the length of the flagstaff that any discrepancies are attributed to errors in measurement. A scientist would say that the limited region of space considered is Euclidean; what he really means is that an abstract model, founded on Euclidean geometry, is in perfect correspondence with this part of the physical universe. The power of mathematics, of course, is much more impressive when it comes to predicting eclipses or designing machines and atomic bombs, but the principle is the same throughout.

We have already referred to the mathematician's habit of framing his ideal world as one in which certain awkward features of the real world do not appear—for example, viscosity (always a troublesome property of a real fluid) is conveniently omitted in the elementary treatment of the lift of an aircraft wing. This process may be regarded in another way. Ideal worlds of this type are, in effect, *limiting cases* of more realistic representations, the argument being, for example, that, since the viscosity of air is very small, we can simplify the analysis, sometimes without sacrificing realism, by ignoring it altogether. This process of 'looking at the limiting case' is used automatically by applied mathematicians as a safeguard against error; having solved a problem involving, say, a variable temperature, the mathematician compares his answer with a previously established result for a constant temperature by verifying that the new result contains the old as a special case. Frictionless pulleys, rigid beams and incompressible fluids are to be re-

---

* For a discussion of what is meant by 'beauty' in mathematics, the reader is referred to G. H. Hardy's essay *A Mathematician's Apology*, from which the quotation which heads this chapter is taken.

garded, in this sense, as limiting cases of real pulleys, beams and fluids, but it must be remembered that the limiting case is an abstraction, and not a special form of a real object.

Among the most interesting and attractive of such systems are those employed in cosmology—the study of the structure of the universe. Despite the great advances which have been made in the design of telescopes and other instruments, our direct knowledge of the physical universe is still very limited. To extend this knowledge we must speculate, and test our conjectures by measurements. In cosmology this is done by the investigation of mathematical concepts called *world-models*. Such concepts, to be significant, must not be mere pipe-dreams or fantasies like those which haunted primitive man, but systems which, in the first instance, exhibit very close correspondence with the world in our immediate vicinity. They are thus necessarily built on a framework of established physical laws, and their prime purpose is to indicate the measurements which should be made to extend such knowledge to the universe as a whole. The entire history of an imaginary universe is written in the postulates of its structure, and the model, to be acceptable, must fit the known facts of the real world at all points and at all times. If, for example, it were found that gravitational forces in a particular world-model would draw all matter together in a huge lump in a relatively short time, the model would have to be abandoned or drastically modified, because we know that in the real world this sort of catastrophe has not happened. Such a world-model is not *incorrect*—it is simply *inappropriate* as a means of exploring the physical world.

The simplest of all world-models was conceived and generally accepted in mediæval times. In this the Earth is supposed to be at the centre of a large hollow sphere, with the stars moving on the surface of the sphere in complicated but unchangeable paths. This is the concept of a *finite universe*. It suited the theological thought of the time (an important feature in those days) because it included heaven and hell and located them with admirable precision. This model is now abandoned except for certain technical purposes, notably navigation, in which it is convenient to regard the stars as points on a celestial sphere. The scheme is adequate for geometrical problems which involve only the relative positions of the more easily recognized heavenly bodies, but is too restricted for the investigation of the physical properties of the universe.

With the birth of natural science in western Europe came a broadening of men's minds and a greater readiness to accept a less conspicuous place in the universe. Newton proposed a model of an *infinite universe*, with an indefinitely large number of massive bodies scattered more or less uniformly throughout space. The geometrical laws assumed for this model were those of Euclid. This system achieved the great triumphs which will always be associated with Newton's name, triumphs so spectacular that the scheme was for long regarded as the ultimate picture of the universe, but towards the end of the nineteenth century doubts began to rise. The astronomer Seeliger, by a simple but convincing argument, showed that the postulated uniformity of distribution of matter could not be reconciled with the famous inverse-square law of gravitation. The difficulty could be evaded if one were prepared to tinker with the law of gravitation by introducing a 'correction' which is negligible except when immense distances are involved, but this is an unsatisfactory way of saving a theory from destruction. Later, other evidence came to hasten the passage of the simple Newtonian world as the complete solution.

At the time when Newton conceived his model, the concept of non-Euclidean geometry had not emerged; but when Einstein began his work in the early years of this century the way had already been cleared by the pure mathematicians. Einstein proposed a world-model for which geometry is non-Euclidean (this, of course, did not mean the abandonment of Euclid's geometry for limited regions), in which space, to use the famous phrase, is 'finite but unbounded'. This particular conception offers no difficulty to a pure mathematician, who is accustomed to calculating in any number of 'dimensions' without having to draw figures (that is, to rely for his arguments on mental images produced by illustrations on paper), but it has created, in some minds, a belief that the Einstein theory is necessarily esoteric. Attempts to explain these concepts in simple language have sometimes increased the confusion. It is doubtful if the non-mathematician is really helped by being asked to think of the four-dimensional counterpart of a flat insect crawling over the finite but unbounded surface of an ordinary sphere. 'Curvature of space-time' is simply the name given to a feature of the mathematical analysis, and is perhaps best left at that.

Einstein's world-model is often described as one in which there is 'matter without motion'. Another world-model, suggested by the Dutch astronomer W. de Sitter, is said to be a universe of 'motion without matter'. These phrases mean that a particle introduced into Einstein's world would remain at rest, whereas in de Sitter's world it would recede from the observer with ever-increasing speed. We can regard both these models as limiting cases, Einstein's representing a possible initial state of the real universe in the very remote past, and de Sitter's the state in the equally remote future, when expansion has proceeded so far that every nebula or group of nebulæ has been deserted by all except the members of its own system.

Models of this kind are not restricted to cosmology. They occur also at the other end of the scale of size, in relation to atoms and electrons. In the nineteenth century, when mechanical invention was at its height, there was a natural tendency to explain everything in terms of mechanisms. The 'billiard ball' model of gas, which gave rise to the kinetic theory of gases, is an example of a highly successful attempt to explain the mass behaviour of the swarms of molecules which make up a gas, and it was natural to look for the same success in other fields. As Eddington put it, 'A man who could make gravitation out of cog-wheels would have been a hero in the Victorian age'. In the present century the tendency has been all the other way. The theories of electrons proposed by Heisenberg, Dirac and Schrödinger are *essentially* mathematical, both in form and content.

What we have learned so far of the universe, both as a whole and in its microstructure, suggests that in neither aspect can it be treated merely as an enlarged or diminished version of the world which we know through our senses. The ultimate secrets of nature are written in a language which we cannot yet read. Mathematics provides a commentary on the text, sometimes a close translation, but in words we can read because they are our own.

2

THE TOOLS OF THE TRADE

The world's a scene of changes, and to be
Constant, in Nature were inconstancy.

COWLEY

Arithmetic and Algebra

WHEN, IN CHILDHOOD, we learn to count, we achieve one of the greatest feats of abstraction of our lives. Ten apples and ten pennies are quite different in their appeal to our senses, and we recognize a common feature only when we lay each penny alongside each apple. We have then demonstrated a unique correspondence between the groups of objects, to which we give the name *number*. Number depends on the concept of a *class*; when two classes have something in common, not possessed by any dissimilar class, we say they have the same number. Thus number is an abstraction, but the concept is accepted easily in childhood and appears difficult only when we attempt to analyse it in later years. Recognition of number seems to be something 'built into' our mental processes, and this may be, perhaps, what Kronecker had in mind when he uttered his famous aphorism, 'God created whole numbers—all else is man's work'.

The manipulation of specific numbers to produce new numbers (which may or may not be associated with concrete objects) is called *elementary arithmetic*. The German mathematician Hilbert once defined mathematics as a game played according to certain rules with meaningless marks on paper (the same thought must often cross the minds of those who correct students' exercises), and in arithmetic the rules are few and simple, the most important being those summed up in the multiplication tables. The numbers of everyday arithmetic are called *real*, but this adjective is a mere label, and is not meant to suggest that they are more genuine and less abstract than any other kinds of numbers.

The manipulation of unspecified numbers is called *algebra*. Like arithmetic, ordinary algebra has a set of simple rules, some of which seem, to the beginner, to be 'self-evident', such as the rule which says that when adding or multiplying, the order in which the quantities are arranged is immaterial. Another rule, that the product of two negative quantities is always positive, seems arbitrary, or even unreasonable. If a child asks *why* minus two multiplied by minus three is plus six, and not minus six, the only answer (which he will not appreciate) is that this is the rule of the game, just as in chess there is no other reason why a pawn moves vertically and takes diagonally.

The object of algebra is to arrange numbers, according to the rules, in different patterns in order to bring to light certain facts of which we were not previously aware. This may seem to be a somewhat trivial reason for a life-study, but in one respect, if no other, algebra is worthy of attention. Pure mathematics is perhaps the only infallible way of extending knowledge (as distinct from mere speculation) beyond the range of direct experience. Consider, for example, one of Fermat's theorems, which asserts that if p is a prime number and n is any whole number not divisible by p, the number $n^{p-1} - 1$ is divisible by p. It is not possible to verify this by actual computation of all cases because the number of primes is infinite, yet no rational being who understands the nature of mathematical proof can entertain any doubt that it is true for all prime numbers, including those as yet undiscovered. Statements which transcend human experience and, by their very nature, are incapable of being tested directly are not uncommon—they appear essentially in sayings or writings of a mystic character—but Fermat's statement differs from all these in that it compels belief now and forever. It is precisely this quality of timelessness and universality which stirs the imagination and constitutes the enduring appeal of mathematics.

In applied mathematics we are rarely concerned with assertions of this type, especially those dealing with primes, since these play no part in physical science. Algebra comes into applied mathematics primarily as a means of facilitating computation and, rather oddly, of bringing to light those features of a pattern which endure even though the rest of the pattern changes. Such features, called *invariants*, play an important part in relativity physics where interest is centred in quantities which have the same measure, no matter what frame of reference is used to describe space. As Eddington puts it, most of the features of a system which are obvious at a first glance are relative; absolute quantities like 'action' and 'entropy' emerge only after much search, and one of the main functions of algebra in applied mathematics is to expose such important clues to the structure of the physical universe.

At the beginning of this chapter Hilbert's celebrated (and perfectly serious) definition of mathematics as a 'game' was quoted, but the game of mathematics differs from others in at least one important respect. In mathematics it often pays to break rules *openly*, because this is one of the surest ways by which mathematics makes progress. The result may be a new algebra, of which there are now many kinds. Naturally, the process is highly sophisticated; not every breaking of rules leads to an interesting result, but to-day we are in a position to state, more or less explicitly, how such generalizations can be made.*

Complex Numbers

In applied mathematics the most important of the new algebras is one which breaks the rule that the square of a number is always positive. We shall need to refer to this algebra in almost every chapter of this book, so a brief sketch of its logical construction is of interest. This is not the historical construction, but a more sophisticated approach which appeals to the mathematician of to-day.

The algebra in question is that of *complex numbers*. Starting with the elements of ordinary algebra—the so-called real numbers—we note that in this system any real number x can be thought of either as a measure of a length OP along a given line (as in plotting graphs) or as the measure of a *displacement*

---

* See, for example, Chapter 3 of *Mathematics, Queen and Servant of Science* by E. T. Bell, New York (1951).

or change of position of a particle from *O* to *P*. If we adopt
the second view (which is natural for an applied mathematician), we need to know three things to make the conception of
a displacement precise. These are: the starting point (*O*),
the distance moved or *magnitude* of *x*, and the *sense* (i.e. whether
backwards or forwards). The last named is conveniently
denoted by the signs + and —. Having defined these
properties, we can use ordinary algebra to solve any problem
involving displacements along a *fixed straight line*. We can add
and subtract displacements and give logical and consistent
intepretations to the operations of multiplication and division
(except division by zero).

Suppose now we wish to extend this work to cover displacements in a *plane*, not only those on a fixed line. We find
it necessary to introduce a *frame of reference*, the most convenient
being a pair of mutually perpendicular axes, *OX*, *OY*, through

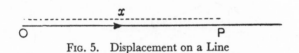

FIG. 5.   Displacement on a Line

a fixed point *O* (as in ordinary graphs) (Fig. 6). Any point
on the plane is then specified by coordinates (*x*, *y*). To
describe a displacement such as *PQ* we must know its starting
point, magnitude and sense, as before, but in addition we
require its *direction* (i.e. the angle which *PQ* makes with either
of the fixed lines *OX*, *OY*). Our problem is now set: can we
invent an algebra (i.e. a logical and consistent set of rules for
calculation), which will enable us to deal with displacements
in a plane just as ordinary algebra enables us to deal with
displacements on a line, in terms of a single quantity? Obviously, this quantity (called a *variable*) must be compounded
of the reference numbers *x* and *y*, just as a molecule is compounded of atoms, and our problem becomes that of finding
how to build up such a number-pair so that, as far as possible,
it will obey the rules of ordinary algebra.

At first there are no difficulties. Equality is defined by
the common-sense requirement that two displacements are
equal if they have the same magnitude, sense and direction.
Thus in Fig. 6 the displacements *PQ* and *OR*, represented by

parallel straight lines of the same length, are equal.   Addition
is straightforward;   the sum of two displacements must be
equal to the displacement which moves the particle to the
same point as that given by the successive application of the
two given displacements.   In Fig. 6 this is illustrated by the
triangle *PQU*, in which *PU* is the sum of the displacements
PQ and *ST*.   We draw the line *QU* parallel and equal to *ST*,
so that *PU* is the diagonal of a parallelogram whose sides are
the given displacements.

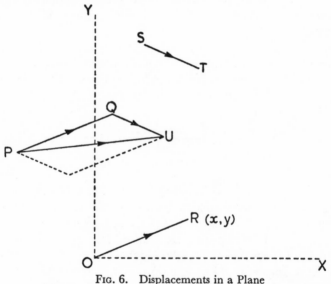

Fig. 6.   Displacements in a Plane

So far we have not suggested an algebraic notation for this
number-pair, or *complex number* as we shall call it henceforth.
We could write it, for example, as $[x, y]$, to distinguish it from
the coordinates of the point (x, y), but instead we anticipate
future developments by introducing the form

$$x + iy$$

where, for the present, i is completely unspecified.   In this
notation addition and subtraction are defined by the rules of
ordinary algebra: for example, if $x + iy$ and $x' + iy'$ are two
complex numbers, their sum is the complex number

$$x + x' + i(y + y')$$

Algebraically, we simply add the constituents of the number-pair, each to each, taking note of sign. It is easy to see that this agrees with the geometrical illustration of addition.

We still have not gone outside ordinary algebra, and the novelty of complex numbers enters when we try to make a rule for multiplication. To investigate this we use geometry again.

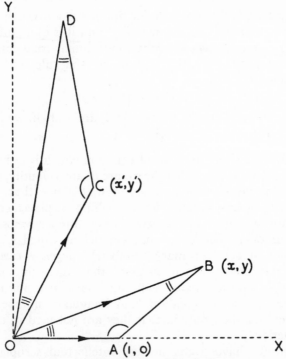

FIG. 7. Multiplication of Displacements

The product of two displacements has a geometrical interpretation in the properties of similar triangles (i.e. triangles which have the same angles, each to each, but different lengths of sides). If (Fig. 7) OAB and OCD are two similar triangles, by ordinary geometry we have $OB/OA = OD/OC$ or

$$OB.OC = OA.OD$$

and this enables us to interpret multiplication of displacements. Suppose OA is 1 unit long; let B be the point (x, y), and C

the point (x', y'). It is easily shown by coordinate geometry that the coordinates of D are $(xx' - yy', \; xy' + yx')$. This suggests that if the displacement OD is to be regarded as the product of OB and OC (OA being a unit displacement), we must have

$$(x + iy) \times (x' + iy') = xx' - yy' + i(xy' + yx')$$

displacement $OB \times$ displacement $OC =$ displacement OD

This, in fact, is the only rule for multiplication which will not lead to a contradiction or absurdity, and it is quite difficult to find unless one is given the hint about similar triangles. Now, if we multiply $x + iy$ by $x' + iy'$ by the rules of ordinary algebra we get

$$xx' + i^2 yy' + i(xy' + yx')$$

This agrees with our proposed rule if, and only if, we put

$$i^2 = -1$$

But this is precisely the kind of equation which is forbidden in ordinary algebra, in which the square of any quantity, whether positive or negative, is positive. We conclude that to make ordinary algebra suitable for describing displacements in a plane we must introduce a new kind of number, whose square is minus one. The older mathematicians called i (and all multiples of i by ordinary numbers) *imaginary numbers*, an unfortunate term which suggests that there is something especially abstract about such numbers. All numbers are creations of the imagination, and imaginary numbers differ from real numbers only in that they are not used in everyday arithmetic.

When we have overcome any intellectual scruples about admitting i into algebra, we find that the rules for calculating with plane displacements are simply those of ordinary algebra with one addition—wherever i^2 occurs we are to replace it by -1. The reason for writing a complex number as $x + iy$ should now be clear; in this form we can operate on it just as we do with any pair of numbers, provided that we remember to replace i^2 by -1. This is an immense advantage, because we do not have to learn the technique of solving algebraic problems all over again when we come to deal with complex numbers.

There is another way of interpreting i, especially useful in applied mathematics. Suppose we represent complex numbers on ordinary graph paper by putting real numbers on the x axis and imaginary numbers on the y axis (Fig. 8). This is called an *Argand diagram*. The number $x + iy$ is represented by the point (x, iy). Multiplication of the real number 2 by i changes it into $2i$, or, in other words, the displacement represented by $(2, 0)$ on the x-axis has been *rotated* through 90° to the point $(0, 2i)$ on the imaginary axis. In this way we may regard i as an *operator* which rotates a displacement

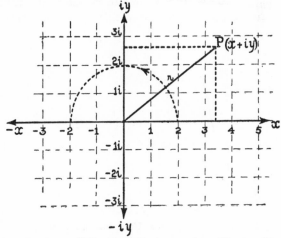

FIG. 8. The Argand Diagram: Interpretation of i as an Operator

through a right angle. A second multiplication by i means another rotation through 90°, bringing the representative point back to the real axis to the position $(-2, 0)$ which is equivalent to reversing the sense of the original displacement. Thus once again i^2 must be interpreted as -1.

The idea of representing a displacement or line by two real numbers linked by an operator of rotation is of such importance in applied mathematics that the algebra of complex numbers is usually developed from this concept. We often replace $x + iy$ by the single letter z, which we call the *complex variable*; z is then a point (P) on the Argand diagram, or *z-plane*. We call x and y the *real and imaginary parts of z*. The distance of z from the origin (OP) is its *modulus*, with the numerical value

$\sqrt{(x^2 + y^2)}$. The direction of the line OP is the angle $(\widehat{XOP})$, whose tangent is y/x; we call this the *amplitude* of z.

As an illustration of the natural way in which complex numbers enter into applied mathematics, consider the use of *vectors*. A physical quantity, like temperature, which can be described by a single number in a system of units (e.g. 60° F.) is called a *scalar*. Other physical quantities, such as velocity or force, require more than one number for their description. Thus velocity has both magnitude and direction (e.g. 30 miles per hour towards the N.W.). Such quantities, called vectors, can be represented by lines of length equal to the magnitude of the quantity, having the same direction and sense as the

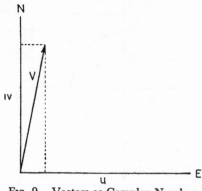

Fig. 9.   Vectors as Complex Numbers

physical quantity. Vectors in a plane are obviously well adapted to representation by complex numbers. The motion of a ship which is steaming north at 10 knots and drifting east with the current at 2 knots could be represented by a combination of velocities $v = 10$ knots to the north and $u = 2$ knots to the east. The vector V, which represents the combined motion, is:

$$V = u + iv = 2 + 10i$$

The quantities u and v are called the *components* of the vector along their respective axes; the components, unlike the vector, depend on the particular axes chosen by the mathematician to facilitate calculation.

Although we have spoken of complex numbers as the constituents of a new algebra, this is exaggerating matters, because the rules are the same as in 'school' algebra. All that has been done is to introduce a new number with one peculiar property. Rotations in *space*, treated in the same way, lead to a genuine new algebra—that of *quaternions*. The distinctive property of rotations in a plane is that the order in which they occur does not matter: the hand of a clock will always end up in the same position as a result of, say, three distinct movements, irrespective of the order in which they occur. We call this the *commutative property* of ordinary algebra; symbolically, this means that $ab = ba$. When we consider three-dimensional space, it is not difficult to imagine sequences of rotations in which order matters a great deal, so that we should expect to find the commutative law broken in algebras which describe such operations. Quaternions, invented by the Irish mathematician W. R. Hamilton to describe rotations in space, are hyper-complex numbers of the form $a + bi + cj + dk$, but instead of the simple rule $i^2 = -1$ of complex numbers, we have a much more complicated system in which $i^2 = j^2 = k^2 = -1$, but $ij = k$ and $ji = -k$, etc.

Quaternions are the oldest example of a non-commutative algebra $(AB \neq BA)$. Hamilton and some of his contemporaries believed that this system ultimately would provide the key to physical science, but the present status of quaternions is shown by the fact that they are not even mentioned in many modern standard texts on mathematical physics.* On the other hand, non-commutative algebras as a class have come to the fore in quantum mechanics. This is not an accident, or merely a device to aid calculation, but the result of a new view of nature. We no longer expect to be able to describe the position and momentum of elementary particles by simple numbers, any more than we can express rotations in this way. Instead, we need combinations of numbers for symbols representing operations, and to do this with the required degree of generality necessitates giving up the commutative law. It also appears that one such algebra is peculiarly fitted to deal with another feature of quantum mechanics, that of indefiniteness or imperfect description, a fact which Whittaker has called 'one of the most remarkable discoveries of the present century'.

* E.g. Jeffreys' *Methods of Mathematical Physics*, Cambridge, 1946.

Hamilton may have been mistaken in his faith in quaternions, but his foresight—one may almost call it courage—in abandoning one of the most cherished and 'self-evident' laws of ordinary algebra has found ample justification in modern science.

We shall not be concerned with these recondite matters in this book. For most purposes in physics the concept of two- or three-dimensional vectors is adequate, but there exist certain important quantities, such as stress in an elastic body, which require more than three numbers for their specification. To deal with these quantities, mathematicians have devised the very general algebra of *tensors*, in which scalars and vectors appear as special cases. Tensor algebra is of prime importance in the development of the world-models of cosmology.

New Geometries

So much for algebra; we turn now to its classroom companion, geometry. In many ways it is much the same story. To see this, consider first the structure of the geometry which has grown from Euclid's *Elements*, the geometry of everyday life. Most schools introduce the subject by exercises in drawing and measurement (that is, by physics), but the true mathematical study begins by naming certain *undefined elements* or *primitive concepts* (points, lines, surfaces) with associated attributes (e.g. a point is said to have 'position, but no magnitude'). There follows a number of self-consistent *axioms* or *dogmatic statements* about the elements, such as 'There can be only one straight line joining two points', 'Every finite straight line has a unique point of bisection', etc. No axiom can be deduced from, or conflicts with, any other, and there never can be any question of 'proving' an axiom. Finally, there is a limited number of *permitted operations*, called *postulates* or *constructions*, such as joining two points by a straight line, producing (prolonging) a straight line to any length and drawing circles of any radius about any point as centre. From these premises certain logical deductions, called *propositions* or *theorems*, are made. The propositions are statements about the descriptive or metrical properties of assemblies of points, lines and surfaces.

In the older books the axioms are described as 'simple principles, the truth of which is so evident that they are accepted without proof'.* This is contrary to the modern view, and to-day even the most unsophisticated student sees that what is acceptable 'without proof' may differ greatly from one person to another. In many ways a system of geometry is like a religion. Every creed begins with certain primitive concepts (gods, demons and other supernatural beings) endowed with certain attributes, benign or evil. The distinctive features of a religion are contained in its dogmas, statements about the elements which the faithful may not question. Certain acts are permitted, others are forbidden. The theology or science of the religion consists of logical deductions from the elements and dogmas, bringing in the favoured activities. Certain political theories, notably Communism, exhibit a similar structure.

The distinctive feature of a religion or political theory, as opposed to a geometry, lies in the fact that its adherents believe their system alone to be true. The dogmas are held to be justified on the grounds that no reasonable (i.e. non-absurd) alternatives exist, and the theory, as a whole, is claimed to provide the only correct interpretation of experience. Up to the beginning of the nineteenth century Euclidean geometry was believed to occupy a similar position in mathematics. The modern view is that geometry is not concerned with truth in this sense and has none of these qualities. The axioms proposed by Euclid are not the only set on which a geometry can be built, and it is now believed that Euclidean geometry is not universally 'true' in the sense of furnishing the correct interpretation of the features of the universe as a whole. This change of outlook is the result, very largely, of the work of the Polish mathematician Nicholas Lobatchevsky, who, in the early years of the nineteenth century, produced the first of many 'heresies', or non-Euclidean geometries, by discarding the famous 'axiom of parallels'. As enunciated by Euclid this reads: 'If a straight line crosses two other straight lines so that the two interior angles on one side are together less than two right angles, the two other straight lines, when produced, will meet on the side on which are the angles whose sum is less than two right angles'. This amounts to the statement, 'The three

* This is a quotation from the opening paragraph of a well-known school text-book of geometry.

angles of a triangle are together equal to two right angles', and there are other equivalent forms. Most people would be inclined to accept the axiom of parallels as a 'self-evident truth', but this particular axiom is the distinctive feature of Euclidean geometry. Lobatchevsky investigated a system which denies this axiom (actually one in which, to use the second form of the axiom, the sum of the three angles of a triangle is *less* than two right angles), and in doing so invented *hyperbolic geometry*. Another type of geometry, in which the three angles of a triangle are together *greater* than two right angles, was proposed by the German mathematician Riemann in 1854—this is called *elliptic geometry*. This geometry has the peculiar property that straight lines, when produced, return on themselves and space is finite. In Einstein's world-model (Chapter 1) a ray of light would circumnavigate the universe, returning to its starting point after a period of the order of 1,000 million years, possibly producing to the observer a ghost star where the original star had been æons before. (We still cannot decide with certainty whether or not some of the faint blurs of light we see at the limit of observation of our telescopes are optical ghosts or not.) The essential point, however, is that we no longer believe that Euclidean geometry is the unique true representation of the physical world, although, to reassure the reader, we can say that the reasons which lead us to doubt that Euclid's geometry is appropriate for surveying the whole universe become evident only when we go far beyond the earth and its immediate companions in space.

To sum up, mathematicians are no respecters of tradition, but incorrigible heretics who are prepared to change their ground and invent new systems of thought as soon as the old is seriously challenged. It must not be thought that this breaking of rules always leads to significant advances. To build a geometry on a new set of axioms demands skill, but not necessarily genius. The chances are, however, that any such new geometry will excite little interest. The true touch of genius appears only when the new algebra or geometry opens up fresh fields of thought, and this is a very rare event.

The Infinitesimal Calculus

The most significant event of the seventeenth century is one which usually receives scant attention from historians. During this turbulent period European scholars, led by Newton in England and by Leibnitz in Germany, revolutionized mathematics by creating its most powerful and subtle instrument—the *infinitesimal calculus*. The rapid advances towards the conquest of nature made in the nineteenth century and in our own time are the direct consequences of this great upheaval in mathematics, which far outweighs in importance battles for sea-power and struggles for new lands.

The invention of the calculus was more than the discovery of a novel device to simplify calculation. It marked the beginning of a new approach to the problem of understanding the physical world. Pre-Newtonian science, especially that of the Greeks, was characterized by attempts to formulate laws which comprised the whole of Nature in a symmetrical, harmonious system. In such theories æsthetic considerations played no small part—certain numbers were thought to be 'perfect', and certain shapes, such as circles and spheres, were specially favoured. The modern method begins by establishing relations between successive phases of a system in restricted regions, and the whole picture is built up from such small elements. The first requirement is therefore a clear understanding of what is meant by an 'infinitesimal' quantity, and the calculus, in brief, is the algebra which enables us to use infinitesimals in calculation with precision and complete confidence.

Before discussing the basic concepts of the calculus (an essential prelude to any attempt to grasp the salient features of modern mathematical physics), some remarks must be made about *rigour* in mathematics. In the early years of the development of the calculus, mathematicians used the new tools freely, without pausing to examine how far their intuitive ideas would bear close examination. Intuition, however, is not a safe guide in mathematics, and the necessity for the establishment of a strict logical framework ultimately became evident. The first step towards real rigour was taken by Bishop Berkeley (1685–1753); he was followed, among others, by Lagrange

(1736–1813), Weierstrass (1815–1897), Riemann (1826–1866) and Cantor (1845–1918), who between them gave the calculus its modern form. How far they succeeded in establishing a completely rigorous basis in not a question to be discussed in this book, but their influence is to be seen in the fact that all modern texts on advanced calculus follow their lead by introducing fundamental concepts, such as continuity, on a strictly *arithmetical* basis. Such formal definitions are difficult to grasp without considerable training in mathematical analysis, and for this reason are usually deferred until a fairly late stage in the education of a mathematician. The exposition given below is descriptive rather than arithmetical, and therefore must not be regarded as completely rigorous in the modern sense, but the ideas are essentially modern versions of those which have prevailed in the calculus from its earliest conception.

We need first to establish a terminology. We have already met the idea of a *variable*, which is no more than an *unspecified number*. The *continuous real variable* x is a quantity which can have any real number as its *value*. From the concept of a variable arises the idea of *functional dependence*. Briefly, we say that y is a *function* of x (written $y = f(x)$) if the value of y changes when x changes or, in other words, when y depends on x. The area of a circle (y) is a function of its radius (x) $(y = \pi x^2)$; in this example we can calculate y for every value of x, but this condition is not necessary for functional dependence. The temperature of the air at a given place depends on the time of day, but meteorology has not yet reached the stage (and is unlikely to do so in the foreseeable future) when we can formulate a precise mathematical relation between temperature and time. Applied mathematics contains many instances of functional relations for which no explicit formula has been found.

The first problem is to make precise what is meant by 'small' and 'large' in mathematics. These words are meaningless without a context, and a little reflection will show that this is so in everyday life. A gift of £1 would probably cause a considerable change in the total wealth of a schoolboy, but it can make no appreciable difference in the annual budget of this country. A million is generally regarded as a large number in population problems, but the addition of a million molecules of oxygen to the air of a moderate-size room could not be detected by the most sensitive instrument. These considerations show that size in mathematics is entirely relative, and from this we are led to the primary conception of an infinitesimal quantity as one which does not exceed the smallest change of which we can take cognizance in our calculations. An infinitesimal quantity thus can be as small as we please (but not zero), and we can decide how large it can be only when we have first decided the magnitude of the least change which will influence the result of the calculation.

To fix ideas, consider the function $y = x^2$, and let δx denote a small change in x.* When x changes to $x + \delta x$ the corresponding change in y is

$$\delta y = (x + \delta x)^2 - x^2 = 2x\delta x + (\delta x)^2$$

If we were told to draw the graph of $y = x^2$ on ordinary ruled paper for values of x between, say, 0 and 5, we would probably decide that there would be no point in calculating y beyond two places of decimals, so that in this case the smallest significant change in y is 0·01. If δy is to be infinitesimal, we must ensure that $2x\delta x + (\delta x)^2$ is never more than 0·01. If we take δx to be less than 0·001, $2x\delta x$ is not greater than 0·01 for all values of x between 0 and 5. The quantity $(\delta x)^2$ is less than 0·000001, far below the limit of detectable change in y. Mathematically, we say that we can always choose a range of values for δx so that $(\delta x)^2$ is *negligible*, and this is a general property of infinitesimals. (The statement applies with even greater force to $(\delta x)^3$ and higher powers.) The whole quantity $2x\delta x + (\delta x)^2$ is an infinitesimal and $2x\delta x$ is called the *principal part*.

This example shows that an infinitesimal is a *variable quantity*, with no definite value except an upper bound which is fixed by a prior choice of what is considered to be the least significant change. In these circumstances no appreciable error is caused by ignoring everything in an infinitesimal except its principal part. This is a fundamental principle in the algebra of infinitesimals.

* The reader will understand that δx does not mean 'δ multiplied by x'—the letter δ has no meaning on its own. δx is a function of x which is restricted to small values, but otherwise undefined.

Consider now the ratio of the change in y and the change in x. In the example given above this is

$$\frac{\delta y}{\delta x} = 2x + \delta x \qquad \ldots(1)$$

This equation is true for *all* changes; the equation

$$\frac{\delta y}{\delta x} = 2x \qquad \ldots(2)$$

is *approximately true* for small changes, but the error in the approximation can be made less than any given number by a suitable choice of δx. Any such choice of δx does not affect the right-hand side of equation (2), so that in this example $2x$ is a quantity which, although derived by a process involving infinitesimals, does not depend on the magnitude of the error we have decided is significant. Mathematically this is expressed by the phrase '$2x$ is the *limit* of $2x + \delta x$ as δx approaches zero' and in symbols by

$$\lim_{\delta x \to 0} (2x + \delta x) = 2x$$

The concept of limits is fundamental in the calculus. Pictorially, a limit is the name written on a signpost, saying that if we proceed along this route we will approach nearer and nearer to the named value, but there is no guarantee that the road will actually pass through the value. (For example, as x approaches the value zero, the quantity $\frac{\sin x}{x}$ approaches nearer and nearer to the value 1 or, in symbols

$$\lim_{x \to 0} \frac{\sin x}{x} = 1$$

but when $x = 0$ it cannot be said that $\frac{\sin x}{x}$ is equal to 1, because $\frac{\sin 0}{0} = \frac{0}{0}$, which is meaningless. On the other hand, $\lim_{x \to 0} \sin x = 0$ and $\sin 0 = 0$, so that a limit may also be a value of the function.) A function which actually passes through its limiting values at all points, irrespective of the direction of approach, is said to be *continuous*; this agrees with the intuitive idea of a continuous function as one which takes all values between its initial and final values—in other words, there are no 'breaks' in its graph. For the present we shall discuss only continuous functions.*

Derivatives and Differentials

Among the various limiting values, that of $\delta y/\delta x$ as δx approaches zero is of such importance that it may be regarded as the central point of the calculus. The limit is called the *derivative of y with respect to x* and is denoted by the symbol y' or $f'(x)$. The above example shows that if $y = x^2$, the derivative y' is $2x$. To extend the idea to any function we write:

$$y' = f'(x) = \lim_{\delta x \to 0} \frac{f(x + \delta x) - f(x)}{\delta x} \qquad \ldots(3)$$

This is a *definition*; $f(x + \delta x) - f(x)$ is what we have called δy, the change in y corresponding to a change of δx in x, and we must divide this by δx, ignore all terms in $(\delta x)^2$, $(\delta x)^3$, ... and find the limit (i.e. a quantity not involving δx) as δx becomes smaller and smaller.

There is another way of looking at this process. The equation

$$\frac{\delta y}{\delta x} = f'(x) \quad \text{or} \quad \delta y = f'(x)\delta x \qquad \ldots(4)$$

is *approximately* true for any infinitesimal δx. Having found $f'(x)$ by the process described above, take any value of δx and call it dx. With dx we associate another infinitesimal dy such that

$$\frac{dy}{dx} = f'(x) \quad \text{or} \quad dy = f'(x)dx \qquad \ldots(5)$$

(We can always do this unless $f'(x)$ is indefinitely large, a case excluded here.) The quantities dx and dy defined in this way are called *differentials*, and for this reason the derivative $f'(x)$ is frequently called the *differential coefficient*.

The introduction of differentials is one of the most brilliant technical devices of the calculus, and it is essential that the

---

* For the formal definitions of a limit and a continuous function see Appendices I and II.

implications of the process described above be clearly understood. We begin with an infinitesimal dx and define another
infinitesimal quantity dy by multiplying dx by $f'(x)$. The
differential dy is *not* the change in y corresponding to dx (except
when $f(x) = x$), but its principal part. It follows that the
ratio of dy and dx is the derivative $f'(x)$, so that *dy/dx is an
ordinary algebraic fraction* with a numerator and denominator
which can be treated as separate entities.

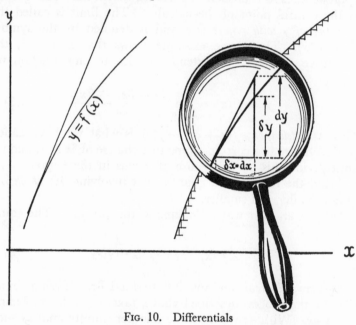

FIG. 10. Differentials

In the *differential calculus* (which, as its name implies, is the
algebra of differentials) we do not use the approximate equation (4), but the *exact* equation (5). In effect, what is done is to
replace a continuous curve $y = f(x)$ by a kind of stairway made
up of elementary (infinitesimal) triangles (Fig. 10). The base
and height of the elementary triangle are dx and dy, respectively,
so that the hypotenuse is $\sqrt{(dx)^2 + (dy)^2}$, which must be the
principal part of the infinitesimal arc of the curve, now replaced by an infinitesimal straight line. It is also clear that
$f'(x) = dy/dx$ must measure the gradient of $f(x)$ or the slope
of the tangent to the curve at (x, y). These are the fundamental

concepts of plane *differential geometry*, or the application of the calculus to plane geometry, in which curves are considered as limiting forms of series of short straight lines.

Differentials do not introduce any radically new ideas—the basic equation is still (4)—but for rapid and easy calculation the concept is almost essential, especially in applied mathematics. There are, however, certain drawbacks, particularly for the learner. Unless the meaning of the notation is explained very carefully, the beginner is apt to find the use of δ and d confusing, often with good reason. Many elementary textbooks begin by defining $\frac{dy}{dx}$ as the symbol for the limit of $\frac{\delta y}{\delta x}$ (without explaining why such a peculiar notation is adopted), together with a warning that dy and dx must not be regarded as separate quantities. Later, the student finds that dy and dx are frequently separated, especially in differential geometry and applied mathematics, which may lead him to think that the calculus is something of a trick—a very clever trick, no doubt, but not what he has learned to expect in mathematics. If the reader has followed the arguments of the previous pages he may be able to take a more charitable view of mathematicians.

The calculus is the algebra of change. In forming the derivative we are finding the expression for the *rate of change of a given function* for any value of the variable. It follows that, in general, the derivative also is a function of the variable. If a quantity is invariable (like 1, 2, 3, . . . , e, π), its derivative must be zero, and conversely. Such quantities are called, very appropriately, *constants*. If $f'(x)$ is a non-zero constant, it follows immediately that $f(x)$ must be of the form $ax + b$, where a and b are constants; $f(x)$ is then said to be a *linear* function of x (the name is derived from the fact that the graph of $y = ax + b$ is a straight line). The rate of change of $f'(x)$ is denoted by $f''(x)$ or d^2y/dx^2 and called the *second derivative*, and so on, for higher derivatives. In the main, applied mathematics is concerned with first and second derivatives only, but derivatives of higher order appear in some problems.

The concept of the rate of change, or growth, of a function and the discovery of rules for its calculation mark the transition from ancient to modern mathematics. The development of the calculus, however, was less sudden than is commonly supposed.

The idea of a limit goes back at least as far as Archimedes, and there are foreshadows of the calculus ('method of indivisibles') before the seventeenth century. The great achievement of Newton and Leibnitz was to resolve these intuitive gropings into a workable scheme. The process of finding a derivative, called *differentiation*, is easier than many parts of 'school' algebra, and doubtless could be taught to quite young children as a pure routine, but the basic concepts are subtle, and demand a certain maturity of thought for their understanding. The logical difficulties of some of the fundamental concepts are not perfectly resolved even to-day, but applied mathematicians, for the most part, are content to leave these matters to specialists in analysis, and in mathematical physics much is taken for granted which might not pass scrutiny on the pure side. As will be shown later, the 'laws' of the physical universe are, in general, naturally expressed in terms of rates of change, so that the calculus is not merely a convenient aid to calculation but an indispensable element in the exploration of the physical world.

Integration

The process which is inverse to differentiation is called *integration*, and technically is more difficult. The primary problem of the differential calculus is to find the rate of change of a given function. In the integral calculus the given function is supposed to represent the rate of growth or the gradient of some unknown function, and the primary problem is to find the *primitive function* which on differentiation yields the given function. Formally, at least, any function can be differentiated by the application of certain simple rules, but there are no such infallible methods of finding the primitive function, and the practical process of integration often demands a high degree of mathematical skill.

These are matters of technique, and as such will not be considered further here. Essentially, integration is a process of summation, as the name implies. This is easily seen, for to recover y (or $f(x)$) from $f'(x)$ means adding together the infinitesimals dy or $f'(x)dx$. Symbolically this is written

$$y = \int dy = \int f'(x)dx$$

the long 's' denoting the operation of integration. The modern view of integration emphasizes this aspect, and in the rigorous treatment the integral of a function is first *defined* as the sum of infinitesimals, and from this definition the property of 'inverse differentiation' is deduced. How this definition arises can be seen from Fig. 11. The area $ABCD$ bounded by the curve $y = f(x)$, the axis of x and the lines $x = a$, $x = b$ can be regarded as the sum of a large number of elementary strips, like $EFGH$, of width δx. Such a strip is intermediate

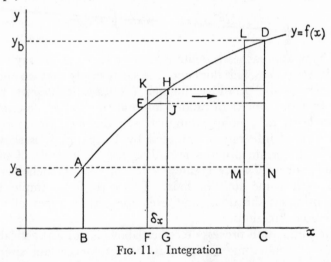

FIG. 11.   Integration

in area between two rectangles, $EFGJ$, of area $y\delta x$ $(EF = y)$, and $KFGH$, of area $(y + \delta y)\delta x$ $(GH = y + \delta y)$. In symbols *

$$y\delta x \leqslant \text{area } EFGH \leqslant (y + \delta y)\delta x$$

If we add together all the elementary areas (using the Greek letter Σ as mathematical shorthand for 'the sum of') we have:

$$\Sigma y\delta x \leqslant \text{area } ABCD \leqslant \Sigma(y + \delta y)\delta x$$

The difference between $y\delta x$ and $(y + \delta y)\delta x$ is $\delta y\delta x$, which is the area of the small rectangle $KEJH$. The sum of all such small rectangles (i.e. $\Sigma\delta y\delta x$) is the rectangle $LMND$ (this can be seen by supposing that each rectangle such as $KEJH$ is moved to the right), which can be made as small as we please by

* The sign '$\leqslant$' means 'does not exceed'.

making δx indefinitely small, since the area $LMND$ is $(y_a - y_b)\delta x$, where y_a and y_b are the values of y for $x = a$, $x = b$ respectively. Hence if $\Sigma y \delta x$ tends to a definite limit as $\delta x \rightarrow 0$, it follows that $\Sigma(y + \delta y)\delta x$ must tend to the same limit, and this limit can be none other than the area $ABCD$. We have thus defined the area between an arc of the curve and the x-axis in terms of a limiting process involving infinitesimals. We call the limit of $\Sigma y \delta x$ the *definite integral* of y (or $f(x)$) with respect to x and write:

$$\text{area } ABCD = \int_{x=a}^{x=b} y dx = \int_a^b f(x) dx$$

If $\Sigma y \delta x$ approaches a definite limit as $\delta x \rightarrow 0$, we say that $f(x)$ is *integrable*. This does not necessarily imply that we know the form of the primitive function (or *indefinite integral*), but only that the process of subdivision and summation described above leads to a unique number. In the modern rigorous treatment of integration (initiated by Riemann) it is shown that, with certain provisos regarding $f(x)$, any mode of subdivision for which the greatest value of δx approaches zero leads to the same limit for $\Sigma y \delta x$. This is so, for example, for all continuous functions, and even for certain types of discontinuous functions.

The summation process described above is not a practicable means of evaluating an integral, except for certain simple functions. Usually, to find a definite integral means finding the primitive function first, and often this is very difficult. If $F(x)$ is the primitive function or indefinite integral of $f(x)$, i.e. if

$$\frac{dF(x)}{dx} = f(x)$$

it follows that

$$\int_a^b f(x) dx = F(b) - F(a)$$

In many problems of applied mathematics $f(x)$ is not given by an explicit mathematical formula, but by a graph, or a series of discrete values. In such instances the definite integral has to be found by a modified summation process or by mechanical means. Integration, of course, is not confined to the problem of areas, but is also used to find lengths of curves and volumes.

In differentiation and integration we have the two fundamental operations of the infinitesimal calculus. We have now to inquire how they are used to extend our knowledge of the physical world.

Differentiation and Integration in Physics

It is not difficult to see why rates of change are of importance in natural science. Consider first the fundamental magnitudes, such as mass, length and time. The notion of the rate of change of length with time is familiar to us as the everyday concept of speed. If the position of a body can be specified by a statement of its distance (x) from an origin at any time (t), we say that x is a function of t, and the velocity (v) of the object is defined as the rate of change of its position with respect to time (i.e. $v = dx/dt$). (It should be noted that it may not always be possible to assign a definite value to x for all t, for example, in dealing with an electron, but here we shall not concern ourselves with these difficultes.) The rate of change of velocity with time is called *acceleration* (retardation is regarded as negative acceleration), or $dv/dt = d^2x/dt^2$. These definitions refer to the *instantaneous* velocity or acceleration, but in practice only the *average* velocity or acceleration over a finite interval of time can be measured. The instantaneous values of the mathematical theory are to be regarded as the limiting forms of average values, as the interval of time is indefinitely reduced. Alternatively, we may know the position of the object at certain times and deduce its position at intermediate times by means of a smooth curve through points on the (x, t) graph. (This process is called *interpolation*.) The velocity at any point or time is then obtained by measuring the slope of the tangent to the curve at the point considered (Fig. 12). It should be noted that in forming the rate of change of length, we have passed from a fundamental to a derived magnitude, and since physical laws involve only derived magnitudes, it is clear that the mathematical expression of some laws will necessarily involve derivatives.

The rates of change of derived magnitudes, such as pressure, may be either with respect to time or length. The latter type of derivative is usually called a *gradient*. Gradients enter into physical laws because throughout nature there is a general tendency towards uniformity (i.e. any local inequalities tend to be smoothed out). For example, it is a common fact of experience that a body which has a high temperature in one region only tends, if left to itself, to become of uniform temperature throughout, as in the example of a poker placed in a fire. Water will flow from one tank to another until the level is the same in both tanks: this is an example of equalization of

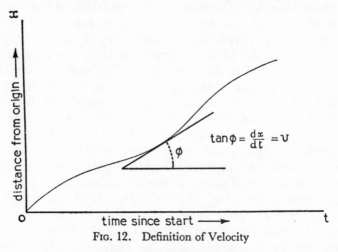

$$\tan \phi = \frac{dx}{dt} = \upsilon$$

Fig. 12.   Definition of Velocity

pressure. If the temperature of a thin wire heated at one end is plotted against distance along the wire, a graph of the type shown in Fig. 13 is obtained. The fundamental mathematical hypothesis of the theory of the conduction of heat is that the rate of flow (or flux) of heat is proportional to the rate at which temperature falls with increasing distance from the heated end, or in symbols

$$\text{flux of heat} \propto \frac{dT}{dx}$$

If the temperature is the same at all points on the wire, $dT/dx = 0$, and there is no net flux of heat.

To proceed further with this concept necessitates an extension of the process of differentiation. So far the discussion has been limited to functions of one variable (e.g. the temperature in the wire was supposed to vary only with distance from one end). If, however, the heat were supplied by an electrical current which was switched off after some time, or was continually changing, the temperature in the rod would depend on both distance and time. We could deduce either the rate of change of temperature with time by reading at different times a thermometer placed at a fixed distance from one end or the rate of change of temperature with distance by a series of thermometers placed along the rod and read at the same time. There is no reason to suppose that these two rates would be

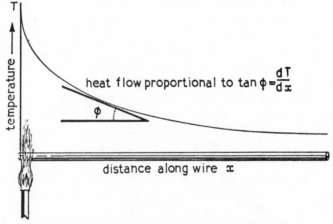

FIG. 13. Definition of Flow of Heat in a Thin Wire

equal, and we must therefore invent a notation which will distinguish between them.

In the complete problem we take into account the fact that temperature changes with position on the wire and with time. If we take a fixed position ($x = $ constant) and consider only the change with time, we can express this variation by what is called in mathematics the *partial* derivative, written $\frac{\partial T}{\partial t}$. The rate of change of temperature with position along the wire is similarly $\frac{\partial T}{\partial x}$. These quantities are calculated by the same rules as for ordinary derivatives, with the proviso that all variables except one are supposed constant during the

operation.   (To make this clear consider a very simple example.
The volume of a cylinder with a circular base of radius r and of
height h is $V = \pi r^2 h$, so that V is a function of r and h.   If r is
allowed to vary, but h is fixed,

$$\frac{\partial V}{\partial r} = 2\pi r h, \quad \text{since} \quad \frac{d}{dr}(\pi r^2) = 2\pi r$$

With r fixed and h varying we have:

$$\frac{\partial V}{\partial h} = \pi r^2, \quad \text{since} \quad \frac{d}{dh}(h) = 1.)$$

In the example of the heated wire the adjective 'thin' means
that any variation of temperature over the cross-section of the
wire is ignored, so that temperature varies spatially in one
direction only.   If the wire were replaced by, say, a massive
metal block it is possible that the fall of temperature would be
different in different directions.   There is, however, always
one direction in which the rate of fall of temperature is a
maximum and zero for all directions perpendicular to this.
The *temperature gradient* in the body is defined as the rate of fall
of temperature in this unique direction, and since temperature gradient has both magnitude and direction, it may be
represented by a vector in three-dimensional space.   This
vector is called the *gradient* and written grad T.   The physical
law for the conduction of heat in the wire, stated above, can
now be expressed in the general form that the flux of heat is
(*a*) proportional to the magnitude of the temperature gradient
and (*b*) always 'down the gradient' (that is, in the direction of
grad T and from higher to lower temperature).   In symbols
we express this law concisely as

$$\text{flux of heat} = -k \text{ grad } T$$

where k is a physical constant, depending only on the nature of
the conducting material, called the *conductivity*.   (The minus
sign is inserted to show that the flow of heat is in the direction
of decreasing temperature.)

To solve a problem in conduction of heat, it is usually convenient to introduce a system of axes, Ox, Oy, Oz.   The rates
of change of temperature along the three axes are

$$\frac{\partial T}{\partial x}, \quad \frac{\partial T}{\partial y}, \quad \frac{\partial T}{\partial z}$$

respectively. These are also the components of the vector grad T along the three axes. The rates of heat flow along the axes are $- k \frac{\partial T}{\partial x}, - k \frac{\partial T}{\partial y}, - k \frac{\partial T}{\partial z}$, respectively.

This example brings out a point of fundamental importance. Exactly as any genuine physical law must be independent of the units used, so must the mathematical formulation of a natural law be independent of any system of axes, for coordinate systems are not part of the physical world, but abstractions imposed on the problem for our own convenience. The direction of grad T is fixed by nature, but we may split up the flow of heat into any components we choose. In general, when we have expressed the unknown quantities of a physical problem in terms of time and the spatial coordinates (i.e. by scalars or vectors), the partial derivatives cannot appear in arbitrary combinations. The aim of the mathematical physicist is to find significant relations between the rates of change of such functions, and every such relation must have a physical meaning independent of the frame of reference (coordinate system) employed. It follows that a true physical law cannot be expressed in terms of one component—all three must be known. Thus $\frac{\partial T}{\partial x}$ by itself has no physical meaning unless the axis of x is in the direction of the temperature gradient, which is simply another way of saying that $\frac{\partial T}{\partial y}$ and $\frac{\partial T}{\partial z}$ are known to be zero. (This is so, for example, in the problem of conduction of heat along the wire.)

This principle acts as a touchstone for the mathematical physicist, by which he can judge, in some measure, the validity of his theories. In dynamical meteorology, for example, the latitude, or position of a place relative to the equator, must enter into the equations of motion of the air because, without restraining forces, a body on a rotating earth inevitably moves towards the equator. The equator is thus part of the real system, and must always appear as a unique line, no matter what system of coordinates is employed. On the other hand, longitude, or position relative to the meridian of Greenwich, cannot affect matters because the position of the prime meridian has been fixed by man, and the indifference of the weather to

the human race is proverbial. The principle appears in a
more striking way in cosmology. Observations of the spectra
of the great nebulæ have been interpreted as evidence that
such systems are receding from us at speeds roughly proportional to their distances from the earth. A world-model in
which our galaxy, alone of all such systems, becomes a unique
'centre of repulsion' (as it were) is rejected by the modern
cosmologist, who, unlike his mediæval counterpart, refuses to
believe that the earth and its neighbours occupy any special
place in the universe. Instead, he is more inclined to accept
theories in which the recession of the nebulæ is universal,
every such system receding from every other, just as any two
spots on an expanding balloon are moving apart, although to an
observer on any spot it appears that he alone is being shunned
by all others.

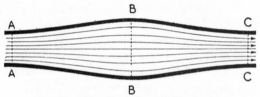

FIG. 14.   The Equation of Continuity

Integration enters into mathematical physics chiefly in the
application of a general law to a special case. This is discussed
in detail later, and here we shall consider integration only in
relation to the formulation of laws. So far we have dealt with
scalars and their vector gradients, but it is also possible to
orm physically significant vectors which are not gradients.
An obvious example arises in the motion of a fluid, in which
velocity is to be regarded as quantity which changes, in magnitude and direction, from point to point and possibly also with
time. One of the first objectives of the mathematical physicist
seeking for a means of describing such complicated systems is
to limit the possible variations of velocity by some overriding
principle which is independent, not only of any coordinate
system, but also of the particular configuration of the problem.
Such a universal principle is that of the conservation of matter.
Suppose we apply this to fluid moving through a pipe of
variable cross-section. For simplicity it is assumed that the

density of the fluid is the same at all points, and also that the velocity of the fluid, measured at a fixed point, does not change with time. The law of conservation of matter then says that the amounts of fluid crossing sections such as *AA*, *BB* or *CC*, are equal. If the velocity does not vary across the pipe, the amount of fluid passing through a cross-section in unit time is simply

(area of cross-section) × (velocity) × (density)

and this must be constant. Since density does not change, it follows that velocity is inversely proportional to the cross-section area, so that to maintain a con-stant discharge the fluid must slow down in going from *AA* to *BB* and speed up in passing from *BB* to *CC*. If, how-ever, the velocity varies over any cross-section (as it does in a real pipe, being least near the wall), it is necessary to split up the cross-section into infinitesi-mal elements of area, multiply each element of area by the local velocity and the density, integrate the product over the cross-section and equate the result to a constant (the discharge), to express the law of conservation of mass.

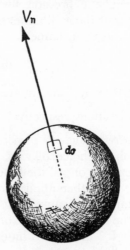

Fig. 15.
Divergence of a Vector

This procedure can be generalized to define an important function which allows the law of conservation of mass to be expressed in terms of an integral or derivatives. We imagine the point in question to be surrounded by an in-finitesimal closed surface (Fig. 15) like a small balloon, the surface of which is divided into small elements $d\sigma$. If V_n denotes the component of the velocity vector perpendicular to the surface, we form the surface integral

$$\int V_n d\sigma$$

which is an infinitesimal of the third order. To obtain a finite quantity we divide by the infinitesimal volume (Ω) and allow Ω to tend to zero. The quantity so formed is a scalar called the *divergence* (div) of V and is written

$$\text{div } V = \lim_{\Omega \to 0} \frac{1}{\Omega} \int V_n d\sigma$$

The physical meaning of the divergence is that when multiplied by the density it represents the rate at which fluid is appearing or disappearing at the point in question. The principle of conservation of mass is that fluid neither appears nor disappears during the motion. Hence div $V = 0$. This is called the *equation of continuity*.

The expression for the divergence as an integral is not particularly illuminating, and there is a simple equivalent expression, namely

$$\text{div } V = \frac{\partial u}{\partial x} + \frac{\partial v}{\partial y} + \frac{\partial w}{\partial z} \qquad \ldots(6)$$

where u, v and w are the components of V in the x, y and z directions, respectively. The derivatives $\frac{\partial u}{\partial x}$, $\frac{\partial v}{\partial y}$ and $\frac{\partial w}{\partial z}$, *by themselves*, have no physical significance, since each is dependent on a choice of axis. In the combination (6) they help to express one of the fundamental laws of nature.

A further application of integration, to the product of the tangential component of the velocity along a closed curve and the infinitesimal arc of the curve, gives the vector known as the *curl*. This also is independent of the coordinate system, and in a fluid measures the amount of rotation in the motion. We shall meet this concept again in Chapter 5, when dealing with mechanical flight. Like the divergence, the curl is usually expressed in terms of the derivatives of the functions.

The use of vectors in mathematical physics often results in a considerable simplification in the form of the equations, but usually, for calculation in a specific problem, it is necessary to locate the vectors in a coordinate system. If the reader has followed the arguments given above concerning the necessity of physical laws being independent of coordinate systems, he will see a reason why vectors are *natural* concepts for the mathematical expression of such laws. In two dimensions, vectors are easily grasped and handled, but vector algebra in three dimensions is more complicated, and considerable practice is necessary before it can be employed with confidence. For this reason we shall not use vectors in any detail in this book, and the expositions will be, for the most part, in the more easily understood systems.

The Mathematical Structure of Classical Physics

Having thus laid out the chief tools of the mathematician, we must now consider how they are used. As explained in Chapter 1, the investigation of a real problem (such as the prediction of the time of sunrise or the final temperature of a body which is being heated or cooled) begins with the formulation of the corresponding ideal problem. The facts of observation must also be cast into the appropriate form, which means that some general principle must be extracted from the measurements. Many of these principles are now regarded as firmly established and are dignified with the title of *laws*. Such laws are of two main kinds:

> (*a*) positive assertions of a mathematical type, generally involving some factor of proportionality called a *physical constant*, and
>
> (*b*) negative assertions, not necessarily of a mathematical kind, which furnish a set of *rules for selection of systems*.

As an example of type (*a*), consider the celebrated law of universal gravitation. Like all such relations, this law originates in certain facts of observation, perhaps of apples falling from a tree or the motion of planets around the sun, from which there emerges the concept that one lump of inanimate matter exerts an influence, which is called the 'force of attraction', on another. In its simplest qualitative form the law of gravitation asserts that the attractive force F depends solely on the masses (m_1 and m_2) of the bodies and the distance (r) separating them, and not on the material of which they are composed or on the nature of the intervening medium. That is

$$F = f(m_1, m_2, r)$$

where f is some unknown function. In the quantitative form given by Newton the law states that two particles, situated anywhere in the universe, attract each other with a force proportional to the product of their masses and inversely proportional to the square of the distance between them. In other words, the unknown function f is given the explicit form

$$f(m_1, m_2, r) = G \frac{m_1 m_2}{r^2}$$

where G, called the constant of gravitation, is a numerical factor which has a precise value independent of the values of m_1, m_2 and r (but of course dependent on the units in which these quantities are measured).*

In the quantitative enunciation, apples, planets and other specific real bodies have disappeared and are replaced by hypothetical entities called 'particles', which have mass but no size. (What is done is to associate a number called 'mass' with the Euclidean 'point', which has position but no magnitude.) The law becomes a statement of the functional dependence of one physical magnitude on certain others and is capable of verification, either directly by very delicate laboratory measurements or indirectly by incorporation into a mathematical theory which aspires to predict, say, the motion of the earth around the sun, and hence the times of sunrise. The law begins with a theory or imaginative hypothesis that the attractive force depends only on mass (and not on colour or temperature or any other physical property of the bodies) and on distance (and not on whether air or water or even 'empty space' separates the bodies), goes on to propose an explicit form, and thus offers itself for experimental test, the criterion being that G must be found always to have the same value, in the terrestrial laboratory or in interstellar space. Most of the fundamental laws of physics are of this type.

The second type of statement, (b), is imported into physics because of the embarrassingly wide range of mathematical possibilities. Without some restrictions, the mathematician could produce all sorts of answers, every one of which would be 'correct' in the sense that no mathematical blunder had been made, and he would be unable to distinguish between them.

---

* G should not be confused with g, which is the *acceleration* experienced by a freely falling body when subject to the gravitational pull of the earth. Thus g is local, but G (which is not an acceleration) is universal.

To avoid this difficulty, mechanical systems in which energy is created from nothing, or destroyed, must be excluded; in heat we need the famous second law of thermodynamics, which says that self-acting systems which allow heat to flow 'against the gradient' without producing any other external effect are 'impossible' (i.e. are excluded from consideration). There are many others. Negative statements of this kind have been called 'postulates of impotence' by Whittaker.* They are essentially different from positive numerical assertions like the law of gravitation in that they cannot be verified by measurement, being not predictions but rules which enable us to select from the infinite variety of possible mathematical developments those which are believed to characterize the real universe. Such selection principles are the nearest approach to 'articles of faith' found in physical science, for they are essentially generalizations drawn from experience, beliefs which may have to be modified, or even abandoned, as science progresses. Whittaker holds that these postulates are fundamental and that a branch of physics can be developed on the same lines as a geometry 'beginning with some *a priori* principles, namely, postulates of impotence and then deriving everything else from them by syllogistic reasoning'.† (This, of course, is a highly sophisticated way of studying a science and so far has been attempted only in thermodynamics.) It should also be noted that the word 'impossible' is characteristic of classical physics, and that in atomic physics the attitude tends to be that denoted by 'too improbable'—in other words, the laws of nature are true in a statistical sense only.

We have now the essentials for building ideal problems—a selection of facts in the form of physical laws and some rules which prevent the structure from becoming 'crazy'. We need also a principle for the incorporation of these into a mathematical problem. In classical physics this is provided by the doctrine of *causality* or determinism. This principle has given rise to innumerable philosophical arguments, but for the mathematician it appears to mean simply that the state of a system at any instant is uniquely linked to, or determined by, the succession of states which immediately precedes it. The

---

* *From Euclid to Eddington*, Cambridge (1949), p. 59.
† Whittaker, *op. cit.*, p. 60.

ultimate objective is to find a mathematical expression which links all states of a system, but since in general this is too difficult to achieve by the straightforward application of the laws of physics, we apply these laws, in the first instance, to *neighbouring* states of the system, those which differ infinitesimally in respect of time and space. In practice this is done by using differentials and forming from them a relation between derivatives and certain given functions (that is, one involving only finite quantities). Such a relation is called a *differential equation,* and the principle of causality is equivalent to the expression of the physical laws as differential equations in which time enters as an independent variable.* The process by which the general expression relating states of the system separated by *finite* intervals of time and space is deduced from the differential equation is called *integration of the equation.*

Differential Equations

There are two main classes of differential equations, *ordinary* and *partial*. In an ordinary equation there is one independent variable, and consequently only ordinary derivatives can appear. The *order* of the equation is the rank of the highest derivative therein. Thus

$$m \frac{d^2x}{dt^2} + k \frac{dx}{dt} + ax = 0 \qquad \ldots(7)$$

is an ordinary differential equation of the second order, in which x is the unknown function, t is the independent variable and m, k and a are (physical) constants. (If m is mass, x is distance and t is time, this equation describes the motion of a pendulum which is subject to friction—the solution is called a 'damped oscillation'.) The equation

$$\frac{\partial T}{\partial t} = \kappa \left(\frac{\partial^2 T}{\partial x^2} + \frac{\partial^2 T}{\partial y^2} + \frac{\partial^2 T}{\partial z^2} \right) \qquad \ldots(8)$$

is a second-order partial differential equation in which the unknown function T depends on t, x, y and z. (If T is temperature, t time, x, y and z spatial coordinates and κ the conductivity, this is the equation of heat conduction.)

* For an example of the process of deriving the differential equation from a physical law see Appendix III.

The equations quoted above have one extremely important common feature—they are *linear*. This means that they do not contain products, squares or higher powers of the unknown function and its derivatives. Most of the differential equations of mathematical physics are, mercifully, linear, but non-linear equations occur fundamentally in one very important branch, namely, fluid motion. The second-order equation

$$\frac{\partial u}{\partial t} + u\frac{\partial u}{\partial x} = v\frac{\partial^2 u}{\partial x^2}$$

arises in fluid mechanics (u is the velocity of the fluid particles and v is the kinematic viscosity). This equation is non-linear because the term $u\dfrac{\partial u}{\partial x} = \dfrac{1}{2}\dfrac{\partial(u^2)}{\partial x}$ involves the square of the unknown velocity.

In general, partial differential equations present much greater difficulties than ordinary equations, but provided that the equations are linear, there exist powerful methods for integrating both types. Non-linear equations, in general, are extremely difficult to handle and in most instances are insoluble by straightforward methods. Many non-linear equations can be dealt with effectively only by high-speed computing machines, and there is little doubt that this technique will be used increasingly in the future.

It would be inappropriate, in a book of this type, to attempt a description of the many methods which have been evolved for integrating differential equations. Instead, we shall turn our attention to the physical meaning of the process. The differential equations of physics are statements of physical laws, and therefore indicate features common to many problems. The same equation can appear in many different guises —for example, equation (7) also represents the motion of a galvanometer needle which is affected by the friction of the air and by the damping action of the electrical currents induced in the surrounding metal. The partial differential equation (8) appears in problems concerning the conduction of heat in a solid, the diffusion of matter in still air and the frictional drag of a surface over which fluid is moving. In mathematical physics the number of problems is immense, but the differential equations are relatively few. Clearly, some additional information is needed to specify any particular problem.

This information is given in the shape of *initial and boundary conditions*. Suppose a rectangular slab of metal is being heated on one side. The subsequent distribution of temperature in the slab is determined partly by equation (8) and partly by the initial temperature of the slab and conditions on the other faces (e.g. these faces might be insulated or kept at fixed temperatures). The differential equation expresses how heat flows, whether it be in a slab of iron, in a raindrop or in the earth, and the initial and boundary conditions specify the particular physical system we wish to study. In all problems these conditions represent particular properties of the solution, but in some instances the conditions are really forecasts of the behaviour of the solution in regions where exact knowledge is denied us. Thus to study the way in which heat is transferred from a hot surface to the surrounding air, it is necessary to bring in the condition that the temperature rise in the air decreases indefinitely with distance from the surface. The solution of the problem amounts to finding a function satisfying the differential equation with exactly this property, and only when this has been accomplished is it possible to assign a definite numerical value to the temperature at any distance from the surface. Boundary conditions of this kind, which involve limiting values, are very often encountered in mathematical physics.

The mathematical significance of the conditions is easily seen. Since the derivative of a constant is zero, it follows that one constant disappears with each differentiation. In the reverse process of integration the constants reappear as arbitrary numbers, and in an ordinary differential equation their values are found from the initial conditions. With partial differential equations there are not only arbitrary constants but arbitrary functions to be determined by the initial and boundary conditions. To solve a problem completely there must be as many physical conditions as there are arbitrary functions (or constants) in the solution of the equation. Otherwise, the problem is indeterminate.

The mathematician obtains his equation and his boundary conditions from physics, but he has something to give in return. First, he can assure the physicist that the solution *exists*. This may seem an unnecessary statement about a formula written on paper, but what is meant is that the physicist may be quite certain that the use of the formula will not involve him in any logical contradictions, such as finding that a physical magnitude at any point may have more than one distinct value at a given time. Second, the solution is *unique*—although the equation has infinitely many solutions, there is only one mathematical expression which satisfies both the equation and the given conditions. Without these assurances it would be impossible to put much faith in mathematics as a guide to the physical universe.

Differential equations are not the only way of formulating physical problems. There are also *integral equations*, in which the unknown function appears inside the integral sign. Usually, a differential equation and its boundary conditions can be recast as one integral equation, and this sometimes has technical advantages.

The essential steps in the solution of a problem in mathematical physics are similar to those in the classical syllogism of logic. The syllogism has the following form:

1. The major premise—a statement of a very general truth, e.g.

   all creatures need oxygen to support life

2. The minor premise—a statement of the relevant features of a particular case, e.g.

   the moon has no atmosphere, rivers or seas

3. The conclusion—the application of the general law to the particular case to bring to light a new fact, e.g.

   there cannot be life on the moon

In a typical problem of mathematical physics, say, one dealing with the distribution of temperature in a slab of metal, the corresponding steps are:

1. The general law, expressed as a differential equation, e.g.

   flux of heat is proportional to the gradient of temperature

   (This is equation (8) in words.)

2. The relevant features of the particular case, e.g.

initially the temperature was the same everywhere; subsequently the surface of the slab is kept at a constant (lower) temperature

(These are the initial and boundary conditions.)

3. The conclusion, or solution of the equation, e.g.

at any subsequent time temperature at any point is given by a certain mathematical expression

This is classical mathematical physics in a nutshell.

Atomistic and Field Theories of Physics

The differential equations of physics fall into two main classes. First, there is the group primarily associated with Newtonian mechanics, in which time is the sole independent variable—these are necessarily ordinary equations. Second, there are the relations which appear in the theories of gravitation, conduction of heat, diffusion, fluid motion, electromagnetism, sound and optics. These are partial differential equations.

It is natural to inquire why there should be this distinction, and the answer is that classical physics uses two distinct kinds of abstractions from nature. If matter is regarded as an aggregate of discrete particles which do not change as they move, any dynamical problem amounts to the determination of the position of each particle as a function of time. This atomistic view of nature forms the basis of Newtonian dynamics, in which the differential equations contain only ordinary derivatives because time is the only independent variable.

The second abstraction brings in one of the most important concepts of classical physics, that of a *field*. Mathematically, the definition of the field of a physical magnitude is simple— it is the region of space in which the magnitude is representable by a single-valued function of time and the spatial coordinates. To see what this means physically, consider a typical problem in aerodynamics, such as the determination of the density changes in a volume of air flowing past a body. The complete description of the flow of a finite volume of air requires an exact knowledge of the motions and interactions of the myriads of molecules which it contains. This is impossible, so that some less ambitious scheme must be adopted. To examine the variation of density means that this physical magnitude must be defined at any point and at any time. Now density is found by dividing the mass of a fluid by its volume, but if we try to apply this to a point by making the volume decrease without limit, a stage is reached with a real fluid when the volume contains an uncertain number of molecules, or perhaps none at all, so that the density at a point could have almost any value at a given time. To avoid logical uncertainties of this type, the mathematical theory ignores the molecular structure of the real fluid, except for its effect on bulk properties. Instead, the average drift of the molecules, which we call the velocity of the stream, is represented by vectors at every point in the space occupied by the fluid. We have then a *field of flow*. Other physical magnitudes, such as density, are considered as *field quantities*, that is, functions of time and space which have definite values at all points and times. The fluid, in effect, becomes a continuous medium, capable of being subdivided without limit into infinitesimal 'fluid particles' which are supposed to have all the properties of fluids in bulk and thus bear no relation to molecules.

(Before this idealization can be used with confidence, we must examine how closely a real fluid behaves like a continuous medium. This amounts to saying that the average distance between molecules and the average time between collisions are both 'negligibly small', or, more precisely, that the linear dimensions of the volume being investigated are very large compared with intermolecular distances, and that 'instantaneous' values of physical magnitudes can be considered as averages over times which are long compared with the intervals between collisions of the molecules. The kinetic theory of gases supplies the data required. At ordinary temperatures and pressures the average distance between the molecules of air is about a hundred-thousandth of an inch, and there are about a thousand million collisions a second. For most problems in aerodynamics the assumption that air is a continuous medium introduces no appreciable error.)

The concept of a field appears in many branches of physics.

The representation of temperature as a function of time and space constitutes a scalar field, and the electrical field near a charged particle or the gravitational field around a massive body are constituted by vectors. The question of what 'carries' the field is irrelevant to the mathematician, however much it may interest the physicist. The essential feature of the mathematical concept is that physical magnitudes have definite values everywhere, and in this type of abstraction physical laws find their expression in partial differential equations because there are now four independent variables, the three space-coordinates and time.

The reader should note that certain field problems, particularly in electricity, involve ordinary differential equations. This is because it is sometimes possible to disregard the space variables by assuming that changes take place instantaneously, and that certain properties, such as capacity and resistance, are 'lumped' at various points. In effect, these are degenerate field problems.

The Equations of Physics and the Laplacian

For generations, mathematicians have devoted much time and energy to the study of the equations of physics. The most famous of these are set out below.

1. *The equations of Newtonian mechanics.* The Newtonian equations are founded on the second law of motion, which defines the relation between the rate of change of momentum of a particle and the applied force. Momentum, the 'quantity of motion', is expressed by the product of mass and velocity. The fundamental equation is

$$\frac{d}{dt}(m\mathsf{V}) = \mathsf{F}$$

where V is the velocity vector, m is the mass and F is the force vector. If the mass does not change during the motion, the relation becomes one between acceleration and force.

2. *The field equations.* These are principally

Laplace's equation

$$\frac{\partial^2 \phi}{\partial x^2} + \frac{\partial^2 \phi}{\partial y^2} + \frac{\partial^2 \phi}{\partial z^2} = 0$$

This is one of the most general equations in physics, first introduced by Laplace in 1787 in a study of Saturn's rings. By giving ϕ different meanings, this equation is made to play an important part in the theories of gravitation, electrostatics, magnetism, current electricity, conduction of heat and hydrodynamics.

The equation of wave motion

$$\frac{\partial^2 \phi}{\partial x^2} + \frac{\partial^2 \phi}{\partial y^2} + \frac{\partial^2 \phi}{\partial z^2} = \frac{1}{a^2} \frac{\partial^2 \phi}{\partial t^2}$$

This equation, which relates to the propagation of undulating disturbances, travelling with velocity a, occurs in the theories of light and other electromagnetic waves (including radio waves), of elastic vibrations and sound.

The equation of conduction of heat.

$$\frac{\partial^2 \phi}{\partial x^2} + \frac{\partial^2 \phi}{\partial y^2} + \frac{\partial^2 \phi}{\partial z^2} = \frac{1}{\kappa} \frac{\partial \phi}{\partial t}$$

If ϕ is temperature and κ conductivity, this is the equation of conduction of heat in a solid body. If ϕ is concentration of matter and κ is diffusivity, this equation relates to diffusion, or mixing of a gas with another.

The Navier–Stokes equations of fluid motion. If u, v and w are component velocities along axes of x, y and z, respectively, and if p is pressure, ρ density and ν the kinematic viscosity, the motion of a fluid is described by three simultaneous non-linear second-order partial differential equations of the type

$$\frac{\partial u}{\partial t} + u \frac{\partial u}{\partial x} + v \frac{\partial u}{\partial y} + w \frac{\partial u}{\partial z} = -\frac{1}{\rho} \frac{\partial p}{\partial x} + \nu \left(\frac{\partial^2 u}{\partial x^2} + \frac{\partial^2 u}{\partial y^2} + \frac{\partial^2 u}{\partial z^2} \right)$$

(To express completely a problem in fluid dynamics, the

equation of continuity (p. 52) and a relation for energy must be added to this set.)

In addition to these, there is the set of relations known as Maxwell's equations of the electromagnetic field, but these will not be considered here.

The reader will have noticed that one group of terms

$$\frac{\partial^2 \phi}{\partial x^2} + \frac{\partial^2 \phi}{\partial y^2} + \frac{\partial^2 \phi}{\partial z^2}$$

appears in all the field equations given above. This group is called the *Laplacian* of ϕ and is usually written $\nabla^2 \phi$. Coincidences of this kind are not accidental, and there are good physical reasons for this repetition.

To see why this is so, we appeal to two results in pure mathematics. In the branch of mathematics known as the calculus of variations it is shown that $\nabla^2 \phi = 0$ (i.e. Laplace's equation) is the condition that the function ϕ should have the minimum mean gradient in space. This expresses the universal tendency, noted on p. 48, to reduce any departure from uniformity to a minimum. Laplace's equation is a means of selecting functions which satisfy this condition.

The second result is of the same nature, but lies a little deeper. Suppose that the function ϕ has a known local value at a certain point in a scalar field. It is readily shown that $\nabla^2 \phi$ *measures the difference between the local and average values of ϕ in an infinitesimal neighbourhood of the point.* This result indicates the physical reason for the form of the wave motion and heat conduction equations. If the deviation of the local value from the average in its neighbourhood is a disturbance from a position of equilibrium (as in the example of a plucked string, taking ϕ to be the displacement), there will appear a restoring force expressed by an acceleration (i.e. by a second derivative with respect to time). This gives the equation of wave motion. If the deviation is a change in temperature, which varies with time (e.g. at a hot spot in a body), the difference between the temperature at the point and its average value nearby will be proportional to the rate of change of temperature with time. This is a restatement of the equation of heat conduction. In the Navier–Stokes equations the group $\nu \nabla^2 u$ expresses the action of the molecular agitation in smoothing away velocity differences by diffusion. The principal equations of mathematical physics are essentially variations on the same theme, that Nature always moves to restore uniformity.*

☆   ☆   ☆

This brief summary has done little more than trace the foundations of the imposing edifice of the classical physics. It is a strange thought that a few symbols, 'meaningless marks on paper', have contributed so much to the shape of life as we know it to-day. A mediæval scholar, awakening in our world, would recognize such symbols as spells, magic formulæ which when properly pronounced confer power over the forces of Nature. In the pages which follow we shall try to show, by examples, how this power has been attained.

* See E. Hopf, "Einführung in die Differentialgleichungen der Physik" (Samm. Göschen, 1933).

3

BALLISTICS, OR NEWTONIAN DYNAMICS IN WAR

. . . every art hath certain rules and principles, without a knowledge of which no man can attain unto a necessary perfection for the practice thereof.

ROBERT NORTON, The Gunner (1628)

Gunnery as an Exact Science

THE SEVENTEENTH CENTURY saw many attempts to apply the calculus to the problem of the motion of projectiles, and ballistics is often cited as one of the earliest examples of technology, the adaptation of the scientific method to purely utilitarian ends. This view is not favoured by modern historians of science.* Until the late nineteenth century, guns and projectiles were so crude, and gunpowder so variable in its composition, that no cannon ball could be relied upon to follow the same path as its predecessor, and the chance of repeating a lucky shot was virtually nil. Nelson's guiding principle, 'to get so close to our Enemies that our shots cannot miss their object', was echoed by most commanders of his day, and long-range shooting was regarded largely as a wasted effort. In such circumstances mathematical theory, however ingenious, contributes nothing to practical gunnery, and it was not until the technical skill of the nineteenth century made the gun into an instrument of precision that the labours of the mathematicians could be put to practical use.

Three conditions must be satisfied before accurate long-range bombardment can be achieved. First, the gunner must know the position of the target in relation to his piece. This is the problem of observation and survey. Second, he must be assured that, within a small margin of error, he can make the projectile follow the same path with successive shots. This means

* See A. R. Hall, *Ballistics in the Seventeenth Century*, Cambridge (1952).

that the propellent is so uniform that the shell always leaves the muzzle with the same velocity and also that the projectile is stable in flight. Third, he must be able to point the gun in a given direction with a very small margin of error—not more than a few minutes of arc. This calls for the solution of the difficult engineering problems of mounting the gun, absorbing the energy of recoil and returning the piece automatically to the same position for the next shot.

In modern artillery these requirements have been met to a degree which makes ballistics not only worth while, but essential, in modern warfare. The subject has grown enormously in the last hundred years. There are now two main divisions: *internal ballistics*, which is concerned with what happens during the passage of the shell along the barrel, and *external ballistics*, which relates to the flight of the projectile through the atmosphere. Internal ballistics is a complex study, involving the thermochemistry of propellents, the behaviour of gases at high pressures and temperatures and the friction of the shell with the wall of the barrel. Here we shall deal only with external ballistics, which in its simplest form is the study of the motion of a massive body, often idealized to a particle (or point-body), acted upon by two forces, gravity and air resistance. In the complete problem the projectile is treated as a rotating body of finite size, affected also by the motion of the earth and meteorological factors. A modern pointed projectile follows a complicated kind of spiral path through the air, but in practice it is often sufficient to concentrate attention on the *mean trajectory*, which is a curve in one plane. In this chapter we shall deal only with the problem of a heavy particle moving along the mean trajectory, as an illustration of the practical use of Newtonian dynamics.

Projectiles *in vacuo*

In the early days of ballistics it was thought that the effects of gravity outweighed all others in importance. If this were so—if gunnery, for example, were practised only on the moon—external ballistics would by now be relegated to the category of solved problems. The motion of a particle in a vacuum, acted upon only by gravity, is no more than an easy exercise in elementary dynamics. In Fig. 16 the shell is supposed to be fired with the initial velocity V_0 at the angle ϕ to the horizontal. Since the only force on the shell is gravity, Newton's fundamental equation is

$$m \frac{dV}{dt} = - W \qquad \ldots(1)$$

where W is the weight or gravitational force. The ratio W/m is the same for all bodies in the earth's gravitational field, and is written g, the acceleration due to gravity ($g = 32$ ft./sec.^{2} = 981 cm./sec.^{2} near the ground). If we impose a set of axes, Ox horizontal and Oy vertical, and assume that the

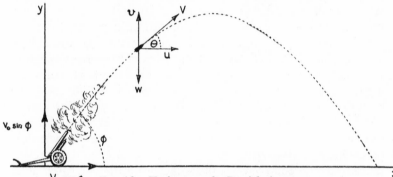

FIG. 16. Trajectory of a Particle *in vacuo*

motion is always in the same (x, y) plane, eq. (1) splits into two component equations

$$\frac{du}{dt} = 0 \qquad \ldots(2)$$

$$\frac{dv}{dt} = - g \qquad \ldots(3)$$

where u and v are component velocities along the axes of x and y respectively. The first of these equations shows that u does not change throughout the motion, and must therefore be equal to its initial value, which is $V_0 \cos \phi$. The second equation gives

$$v = - gt + A$$

where A is a constant, to be determined by the initial condition. If the gun is fired at $t = 0$, A must be the value of v at $t = 0$ or $V_0 \sin \phi$. Hence

$$v = V_0 \sin \phi - gt$$

The velocity $V = \sqrt{(u^2 + v^2)}$ is thus determined at any time t. To find the trajectory we must know x and y as functions of t. Now $u = dx/dt$ and $v = dy/dt$. Hence

$$\frac{dx}{dt} = V_0 \cos \phi \quad \text{or} \quad x = V_0 t \cos \phi + B \qquad ...(4)$$

where B is another constant. Since $x = 0$ at $t = 0$ it follows that $B = 0$ or $x = V_0 t \cos \phi$. Similarly,

$$\frac{dy}{dt} = V_0 \sin \phi - gt \quad \text{or} \quad y = V_0 t \sin \phi - \tfrac{1}{2}gt^2 + C \quad (5)$$

where C is a constant of integration. Since $y = 0$ when $t = 0$, it follows that $C = 0$ and $y = V_0 t \sin \phi - \tfrac{1}{2}gt^2$. Finally, we can eliminate t from the equations and get

$$y = x \tan \phi - \frac{gx^2}{2V_0^2 \cos^2 \phi}$$

which is the equation of a parabola. It follows without difficulty that the *range*, or the second point at which $y = 0$, is given by

$$x = \frac{V_0^2}{g} \sin 2\phi$$

This is a maximum when $\phi = 45°$. The maximum range is V_0^2/g.

This problem shows, in a very simple manner, how the initial conditions enter into Newtonian dynamics. Without these conditions the trajectory could be anywhere in space. The initial conditions not only pin the trajectory to earth but also select out of the infinity of possible trajectories the unique path of a shell fired with muzzle velocity V_0 at an angle ϕ to the horizontal. The same principles hold for more complicated problems.

The Resistance of the Air

A simple example shows the importance of air resistance in ballistics. A body projected from the earth at 2700 feet per second at an angle of 30° to the horizontal would have a horizontal range of about 65,000 yards if there were no atmosphere. A well-designed shell, fired with the same initial velocity and elevation, hardly achieves 30,000 yards range in still air. In other words, gravity and air resistance are about equally effective in limiting the flight of a projectile.

Almost the whole difficulty of external ballistics springs from two causes: first, the relation between air resistance and the speed and shape of a projectile is extremely complicated and imperfectly understood even to-day, and second, the density of the air, which affects resistance directly, varies with height. The first of these factors is the more troublesome. At moderate speeds (below 800 feet per second) the *drag*, or direct resistance opposing the motion of a projectile, is tolerably well expressed by a law first enunciated by Newton, namely

$$\text{drag} = (\text{factor of proportionality}) \times \rho d^2 V^2 = \kappa \rho d^2 V^2$$

where ρ is the density of the air, d the diameter of the shell and V its velocity. The factor of proportionality κ is called the *drag coefficient*.*

If the Newtonian law held at all speeds, external ballistics would present few serious problems to the mathematician. However, gunners have been insatiable in their demands for higher projectile velocities, and when these approach the speed of sound-waves in air (about 1100 feet per second in the lower layers of the atmosphere) the Newtonian law breaks down (i.e. the drag coefficient is no longer independent of V). The reasons for this are discussed in Chapters 4 and 5, and here it suffices to say that at these high speeds the whole pattern of flow around the body is changed because the air is squeezed into a gas of higher density by the passage of the projectile. Fig. 17 illustrates the change in drag coefficient as velocity increases. Just before the speed of sound is reached the drag coefficient takes a sudden, almost dramatic, leap upwards and then declines, but more slowly, in the supersonic region. (An increase in drag coefficient is simply another way of saying that resistance increases more rapidly than the square of the velocity.) The region of rapid change of drag coefficient is popularly called the 'sound barrier', but the technical term *transonic region* is to be preferred. The word 'barrier' suggests that resistance falls when the speed of sound is passed, but this is not so; despite the steady fall of the drag coefficient in the supersonic region, the resistance continues to increase rapidly with the increasing velocity.

* In aerodynamics the drag coefficient is often represented by the symbol C_D defined in a slightly different way by the equation: drag $= \frac{1}{2} C_D \rho d^2 V^2$.

Curves of the type of Fig. 17 have been obtained by ballisticians in many countries. The peculiar properties of the transonic region seem to have been discovered about 1740 by the English ballistician Benjamin Robins, who, in his *New Principles of Gunnery*, stated that for velocities greater than about 800 feet per second 'the resisting power of the medium is augmented to near three times the quantity assigned by [Newton's] theory'. Robins, like his immediate successors Hutton (England) and Didion (France), used the *ballistic pendulum* to measure projectile velocities. In this a massive

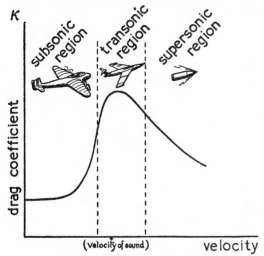

Fig. 17.   Air Resistance and Speed

block of wood, arranged to swing as a pendulum, is placed in the path of the projectile at a known distance from the muzzle. The observer notes the arc of the oscillation set up by the impact of the shot, and from this it is an easy matter to calculate the striking velocity.

The determination of the retardation of the projectile means that velocity must be measured at two or more points on the trajectory, and this presented serious difficulties in the eighteenth century, when powder charges could not be trusted to give the same muzzle velocity on different occasions. Robins overcame this difficulty by mounting the gun on another pendulum and adjusting the powder charge to give tolerably

consistent muzzle velocities. The modern method of deter-
mining retardation was evolved by the Rev. Francis Bashforth,
Professor of Mathematics in what is now the Royal Military
College of Science, in 1864, when pointed projectiles were first
coming into common use. At this time, to quote Bashforth,
'there was no satisfactory work on ballistics and no experiments
made with elongated shot which were consistent among themselves and therefore deserving of confidence'. He rejected the
ballistic pendulum as an instrument of precision, and despite
lack of support from the Ordnance Committee of his day,

FIG. 18.   Bashforth's Method of Measuring Projectile Velocities

devised a scheme whereby measurements of velocity could be
made over different stretches of the same trajectory. The
essentials of his method are shown in Fig. 18. A series of
'screens' (frames criss-crossed by thin copper wire), is placed
in the path of the projectile and connected to an ingenious
electrical chronograph.*   As the projectile passes through the
screens it breaks the circuits in succession, allowing the times
of passage to be recorded on the drum of the chronograph.
From these times the mean velocity over the measured intervals
is calculated, then the retardation and finally the resistance.

* This apparatus is still in good working order and was exhibited at the Physical
Society's exhibition in 1948.

Bashforth continued his experiments until 1880, and obtained results for muzzle velocities as high as 2780 feet per second, nearly two-and-a-half times the speed of sound. In doing so he placed ballistics, for the first time, on a secure footing. Present-day methods of investigating the resistance laws of projectiles are essentially those of Bashforth, and a recent examination, by the Royal Military College of Science, of his 1869 results for cannon-balls showed that these are very little inferior, if at all, to data obtained within the last few years, using all the refinements of modern electronic technique in specially designed ballistic laboratories. When it is considered that Bashforth made his measurements in the open air, at a time when electrical instruments were still something of a novelty, the excellence of the work of this neglected genius becomes evident.

Bashforth was followed by Mayevski (Russia, 1869), Krupp (Germany, 1879–96), Hojel (Holland, 1884) and by the 'Gavre Commission' in France (1884). Most of the investigators attempted to express resistance by various powers of the velocity, but the modern fashion was set when the Gavre Commission produced its results in the form of a tabular statement of the resistance for equally spaced values of the velocity. In all instances the results were in substantial agreement with those of Bashforth, the differences being mainly those arising from variations in shell design.*

The effect of shape on resistance is not easily defined. At the higher velocities the contour of the head is very important, but the shape of the tail also exerts some influence. The determination of the drag coefficient from firing trials is such a lengthy and expensive process that it cannot be attempted for all projectiles in common use, and the usual practice is to select one shape of head as a standard, the resistance laws for other shapes being expressed by empirical 'correction factors' to the standard law. Wind tunnels are sometimes used to investigate the effects of changes in shape, but as yet the amount of information which can be obtained in this way is limited by technical difficulties.

---

* Bashforth (*An Historical Sketch, etc.*, p. 23) suggested that some of the laws given by Mayevski and Krupp were based entirely on his measurements. See Appendix IV.

Gunnery in the Real Atmosphere

So far we have seen how to calculate the path of a shell in a vacuum and also how data have been collected on the resistance which the shell experiences in moving through the air. The resistance data, however, are for the layers of air near the surface of the earth, whereas modern projectiles frequently rise many miles above the surface. At these heights the density of the air is much less than in the lower layers, and resistance is correspondingly reduced. Variation of density with height, however, is not the only factor to be considered—density, being dependent on pressure and temperature, changes also with season and time of day. Finally, a projectile encounters winds of variable speed and direction in its passage through a great thickness of atmosphere, so that, all told, the effects of meteorological factors can be very considerable.

These difficulties are overcome by defining, in the first place, a hypothetical *standard ballistic atmosphere*, in which the fall of density with height is expressed by a specific mathematical function. Further, no winds blow in this ideal atmosphere. In the British standard atmosphere the constants are typical of average conditions in this country during a pleasant day in early summer or autumn, but the choice of the constants really is immaterial. The gunner relies on the meteorologist to supply him with information about the state of the real atmosphere during the period of shooting, and this is applied in the form of 'corrections' to the standard atmosphere.

With the resistance law and the standard atmosphere, the ballistician has everything he needs to begin the calculation of the trajectory. Experience has shown, however, that it is convenient to make a slight rearrangement of the terms in order to express the efficiency with which a given projectile is able to overcome the resistance of the air. It is a matter of common experience that certain objects can be thrown farther than others. A cricket ball (diameter $2\frac{3}{4}$ in.) presents to the air a disc of area more than three times that of a table-tennis ball (diameter $1\frac{1}{2}$ in.), but the table-tennis ball can be thrown only about one-sixth as far as the cricket ball. The reason is that the mass of the cricket ball is nearly a hundred times that of the table-tennis ball, so that the thrower is able to impart a greater momentum to the cricket ball. In symbols, if r is the retardation due to air resistance R, and m is the mass of the projectile, Newton's second law of motion gives

$$R = mr$$

When resistance is expressed by $R = \kappa \rho d^2 V^2$, where κ is the drag coefficient, we get

$$r = \frac{R}{m} = \kappa \rho V^2 \left(\frac{d^2}{m}\right)$$

The quantity m/d^2 is called the *standard ballistic coefficient* * of the projectile and denoted by the symbol C_0. Thus

$$r = \frac{\kappa \rho V^2}{C_0} \qquad \qquad ...(6)$$

and the greater C_0, the smaller is the retardation and the greater the ranging power of the shell. (In the example given above, the cricket ball has a ballistic coefficient nearly thirty times that of the table-tennis ball.)

The Calculation of Trajectories

The practical gunner uses the results of the mathematical theory in the form of *range tables*, which are tabular statements of horizontal range, time of flight, striking velocity and angle of descent at impact, for various angles of elevation. The designer supplies the ballistician with the muzzle velocity and the standard ballistic coefficient of the shell.

The exact methods used for calculating trajectories vary from country to country, and the details are rarely disclosed. It is safe to say, however, that the essential principles must be the same everywhere, because there is only one basic method of solving the problem completely. We shall give here an outline of the British method, as evolved by a group of eminent mathematicians during the war of 1914–18.

The *ballistic elements* at a point on the plane trajectory are: the position coordinates x, y, the time since firing, t, the inclination of the shell to the horizontal, θ, and the velocity along the trajectory, V. The main problem of external ballistics is to express any four of these elements in terms of the

---

* Actually, the British standard ballistic coefficient includes a factor expressing the effect of shape.

remaining element. Thus the designer may wish to know the height of the vertex (or highest point of the trajectory)—this means finding the value of y for which $\theta = 0$—or he may require the velocity at impact, which means a knowledge of V when $y = 0$, and so on.

The first step is to write down the equations of motion in component form, and this presents no difficulties. The first two equations are simply eqs. (2) and (3) of the *in vacuo* problem with the retardations due to air resistance included. They are

$$\frac{du}{dt} = \frac{d^2x}{dt^2} = - r \cos \theta \qquad \ldots(7)$$

$$\frac{dv}{dt} = \frac{d^2y}{dt^2} = - g - r \sin \theta \qquad \ldots(8)$$

The two remaining equations express the resolution of V along the x and y axes and therefore have the same form as *in vacuo*, namely

$$u = \frac{dx}{dt} = V \cos \theta \qquad \ldots(9)$$

$$v = \frac{dy}{dt} = V \sin \theta \qquad \ldots(10)$$

When r is expressed by eq. (6) these equations can be rearranged as follows, using the inclination θ as the independent variable.*

$$\frac{d(V \cos \theta)}{d\theta} = \frac{\kappa \rho V^3}{C_0 g} \qquad \ldots(11)$$

$$\frac{dx}{d\theta} = - \frac{V^2}{g} \qquad \ldots(12)$$

$$\frac{dy}{d\theta} = - \frac{V^2}{g} \tan \theta \qquad \ldots(13)$$

$$\frac{dt}{d\theta} = - \frac{V}{g \cos \theta} \qquad \ldots(14)$$

If we could solve eq. (11) (i.e. find V in terms of θ), we could solve all four equations, and we would then have the elements x, y, t and V expressed in terms of θ. But we cannot do this immediately because κ and ρ vary from instant to instant. Eq. (11), sometimes called the *hodograph equation*, is the key to the whole problem.

---

* The details are given in Appendix V.

Before dealing with the full problem, let us look at a very well-known and widely used approximate method devised by the Italian ballistician Siacci in 1880. Siacci's solution applies to very flat trajectories, in which the inclination θ does not exceed $4°$. Now $\cos 4° = 0.9976$, which may be regarded as unity without serious error, and because the projectile cannot rise to great heights, change of air density with height may be disregarded, so that ρ becomes a constant. In this system eq. (11), when inverted, becomes, in differential form,

$$d\theta = \frac{C_0 g}{\kappa \rho V^3}\, dV \qquad \text{...(11a)}$$

On integration this becomes

$$\int d\theta = \frac{C_0 g}{\rho} \int \frac{dV}{\kappa V^3}$$

If the shell is fired at an elevation ϕ with muzzle velocity V_0, it follows that $\theta = \phi$ when $V = V_0$. Hence

$$\int_\phi^\theta d\theta = \theta - \phi = \frac{C_0 g}{\rho} \int_{V_0}^V \frac{dV}{\kappa V^3}$$

Since κ is known numerically for all values of V, the integral on the right-hand side can be evaluated by a numerical process (p. 46) and expressed as a function of V_0 and V. This function is called *Siacci's Inclination Function*, and by means of special tables θ can be calculated for any velocity V. Similarly, the remaining equations can be treated to give the horizontal distance x, the height y and the time t as functions of velocity, by introducing more tabulated functions called the *Space Function*, the *Altitude Function* and the *Time Function*, all of which are integrals involving κ and V. It is then a simple matter to calculate any element in terms of any other,* and the ballistic problem is solved for flat fire.

Siacci's method was later extended to trajectories for which θ does not exceed $10°$, but since this introduces no essentially new features and is a little complicated, the solution will not be described here. Instead, we shall pass on to the method which applies to any trajectory, especially those in which the approximation $\cos \theta = 1$ cannot be used.

The method is known as *step-by-step integration* or *solution*

---

* In practice this is done by the use of certain auxiliary quantities known as *Secondary Ballistic Functions*.

by small arcs. The essential principle is very simple. The trajectory is divided up into a fairly large number of arcs, and the elements at the end of a short interval are calculated by approximate methods from their values at the beginning of the interval. These end-values then become the starting values for the next interval and so on, until the trajectory is completed. The size of the interval (and therefore the number of arcs used) is defined by the degree of accuracy required, the criterion being that the accumulated error at the final stage must be acceptably small. A characteristic feature of the work is that a certain amount of guessing has to be employed, but the validity of each guess is verified later, and if the result is unsatisfactory, a fresh start is made. Much depends on the experience of the computer and the amount of prior knowledge he has—his task is much simplified if he is working on a trajectory for which the initial conditions (V_0 and ϕ) and the ballistic coefficient are not too dissimilar to those for a previously computed trajectory. The number of arcs is often quite large (perhaps 50), and the time taken by an average experienced computer to complete a calculation is at least a week and often more.

The British method was evolved during the 1914–18 war, mainly by Dr. A. T. Doodson, F.R.S.* Suppose that at one point on the trajectory we know all the elements x, y, t, θ and V. (There is always one such point—namely, $x = y = 0$—where, for $t = 0$, $\theta = \phi$ and $V = V_0$.) One element is chosen as the *independent variable*; in the British system this is time (t), but in theory any element could be selected. The calculation of the other elements is carried out for equal intervals of time—say 1 second—except perhaps in the transonic region, where the intervals may have to be as short as $\frac{1}{4}$ second because of the rapid changes in the drag coefficient. This means that the four ballistic equations have to be rearranged in terms of t as the independent variable, so that all integrations are in terms of t. Now it can be shown that any integral can be expressed in terms of a mean value, i.e.

$$\int_{t_0}^{t_1} \mathcal{Z}dt = (t_1 - t_0)\bar{\mathcal{Z}}$$

* Now Director of the Liverpool Observatory and Tidal Institute.

where $\mathcal{Z}$ represents one of the four remaining elements (x, y, θ and V) and $\bar{\mathcal{Z}}$ is some mean value of $\mathcal{Z}$. The problem amounts to finding a suitable approximate value for $\bar{\mathcal{Z}}$, but we can always carry out the integration to the requisite degree of accuracy by making $t_1 - t_0$ small enough, so that $\mathcal{Z}$ cannot change very much during the interval. However, if the basic interval $t_1 - t_0$ is made very small, the number of arcs becomes prohibitively large and the computation is then extremely laborious. Clearly, this is a case for the 'happy mean'!

The simplest choice for $\bar{\mathcal{Z}}$ is the arithmetic mean, or average of the values $\mathcal{Z}_0$ and $\mathcal{Z}_1$ which $\mathcal{Z}$ takes at the beginning and end of the arc. Then *

$$\int_{t_0}^{t_1} \mathcal{Z} dt \simeq (t_1 - t_0)\left(\frac{\mathcal{Z}_1 + \mathcal{Z}_0}{2}\right)$$

This expression is exact when $\mathcal{Z}$ is a linear function of t, and in many trajectory calculations this approximation is adequate when $t_1 - t_0$ is 1 second. Sometimes more elaborate formulæ, variants of Simpson's Rule, are used.

The first step is to rearrange the ballistic equations with time as the independent variable. The drag coefficient κ is written as $\kappa(V)$, to indicate dependence on velocity, and the density ρ is replaced by a standard function of height, $f(y)$. The equations are then formally integrated in terms of mean values over the interval t_0 to t_1. (The suffixes 0 and 1 denote the values of a quantity at the beginning and end of the interval, respectively, and a bar over a quantity indicates its mean value in the interval.)

(exact) differential equations	(approximate) integrated equations
$d\left(\dfrac{1}{u}\right) = \dfrac{\kappa(V)\sec\theta}{C_0 f(y)}\,dt$	$\dfrac{1}{u_1} - \dfrac{1}{u_0} = \dfrac{\kappa(\bar{V})\sec\bar{\theta}}{C_0 f(\bar{y})}(t_1 - t_0)$ ...(15)
$d(\tan\theta) = -\dfrac{g}{u}\,dt$	$\tan\theta_0 - \tan\theta_1 = \dfrac{g}{\bar{u}}(t_1 - t_0)$ ...(16)
$dx = u\,dt$	$x_1 - x_0 = \bar{u}(t_1 - t_0)$
$dy = u\tan\theta\,dt$	$y_1 - y_0 = \bar{u}\tan\bar{\theta}(t_1 - t_0)$

The table overleaf shows how the work is arranged. This example refers to an obsolete type of shell fired at $\phi = 40°$, but the principles are the same for all trajectories.†

* The symbol $\simeq$ means 'is approximately'.

† The example is taken from an official publication relating to methods developed in the 1914–18 war, and is published by permission of the Controller of Her Britannic Majesty's Stationery Office (Crown copyright reserved).

EXAMPLE OF TRAJECTORY CALCULATION BY SMALL ARCS

Doodson Even-Time Method; $t_1 - t_0 = 1$ second

Line number	Quantity	Time Since Firing $t = 25$ sec.	$t = 26$ sec.	Method
1	$\frac{1}{2}(y_1 - y_0)$	$-20\cdot40$	$-36\cdot30$	*Estimated*
2	y_0	$9526\cdot67$	$9485\cdot88$	**From line 28 of previous column**
3	y	$9506\cdot27$	$9449\cdot58$	Calculated
4	$C_0 f(\bar{y})/(t_1 - t_0)$	$6\cdot5219$	$6\cdot5099$	Calculated
5	$\tan \bar{\theta}$	$-0\cdot051090$	$-0\cdot091500$	*Estimated*
6	$\bar{V}$	$800\cdot15$	$797\cdot37$	*Estimated*
7	$\kappa(\bar{V})$	$0\cdot51337$	$0\cdot51409$	From tables
8	$\sec \bar{\theta}$	$1\cdot00130$	$1\cdot00418$	Calculated
9	$\dfrac{1}{u_1} - \dfrac{1}{u_0}$	$0\cdot000007882$	$0\cdot000007930$	Calculated
10	$1/u_0$	$0\cdot001247501$	$0\cdot001255383$	**From line 11 of previous column**
11	$1/u_1$	$0\cdot001255383$	$0\cdot001263313$	Calculated
12	u_1	$796\cdot570$	$791\cdot569$	Calculated
13	3rd difference of u_1	$-0\cdot011$	$-0\cdot011$	Calculated from 3 arcs. Check
14	u_0	$801\cdot603$	$796\cdot570$	**From line 12 of previous column**
15	$2\bar{u} = u_0 + u_1$	$1598\cdot173$	$1588\cdot139$	Calculated
16	$\bar{V} = \bar{u} \sec \bar{\theta}$	$800\cdot12$	$797\cdot39$	Calculated and checked with 6
17	$\tan \theta_0 - \tan \theta_1$	$0\cdot040284$	$0\cdot040538$	Calculated
18	$\tan \theta_0$	$-0\cdot030947$	$-0\cdot071231$	**From line 19 of previous column**
19	$\tan \theta_1$	$-0\cdot071231$	$-0\cdot111769$	Calculated
20	3rd difference of $\tan \theta$	0	0	Calculated from 3 arcs. Check
21	$x_1 - x_0$	$799\cdot08$	$794\cdot07$	Calculated
22	x_0	$21459\cdot11$	$22258\cdot19$	**From line 23 of previous column**
23	x_1	$22258\cdot19$	$23052\cdot26$	Calculated
24	3rd difference of x	$0\cdot04$	$0\cdot05$	Calculated from 3 arcs. Check
25	$u_1 \tan \theta_1$	$-56\cdot740$	$-88\cdot473$	Calculated
26	$v_1 - y_0$	$-40\cdot79$	$-72\cdot62$	Calculated
27	y_0	$9526\cdot67$	$9485\cdot88$	**From line 28 of previous column**
28	y_1	$9485\cdot88$	$9413\cdot26$	Calculated
29	3rd difference of y	$0\cdot21$	$0\cdot20$	Calculated from 3 arcs. Check

We suppose that at least three arcs have been calculated. The computer begins by entering all the initial values—these are quantities with suffix 0—which, of course, are the end values of the previous arc (those with suffix 1). In this way he fills in lines 2, 10, 14, 18, 22 and 27 (printed in **heavy** type in the example). Next, he *estimates* the value of $y_1 - y_0$ (the difference in height) from inspection of the trend in previous arcs (sometimes using for this purpose an extrapolation formula); this enables him to fill in line 1. He does the same for $\bar{V}$, using values of $\bar{V}$ in line 16 in the previous arcs as a guide, but this estimate is entered in line 6 only. The quantity $\tan \theta_0 - \tan \theta_1$ is estimated and used, with $\tan \theta_0$, to give an estimate for the mean value $\tan \bar{\theta} = \frac{1}{2} (\tan \theta_0 + \tan \theta_1)$ in line 5. (In the example all estimations are printed in *italics*.)

This completes the preliminary stages; the computer now fills in $\bar{y}$ in line 3 from the values of y_0 and y_1, and line 4 is completed from the given value of the ballistic coefficient and the values of density against height in the tables of the standard ballistic atmosphere. Line 7 is filled by reference to the tabulated values of the drag coefficient, and in line 8, $\sec \bar{\theta}$ is computed from $\tan \bar{\theta}$ (line 5). Line 9 can now be filled in from solution (15) of the first ballistic equation, since κ, $\sec \bar{\theta}$ C_0 and $f(\bar{y})$ are all known; the result is added to $1/u_0$ to give $1/u_1$ in line 11, and hence u_1 in line 12.

At this stage it is necessary to test the validity of the estimates. The computer adds u_1 (line 12) to u_0 (line 14) to give $2u$ (line 15), and hence $\bar{V} = u \sec \bar{\theta}$ in line 16. This value of $\bar{V}$ is compared with that written down in line 6. If the two values agree within, say, 0·1, the work may proceed, but if not, new estimates must be tried. (In the example, the two values are 800·15 and 800·12, difference 0·03, which is satisfactory.) A further verification is obtained by comparing the values of u_1 in three consecutive arcs by computing what is called the 'third difference' (line 13). If this is small, the values of u, the horizontal component of V, are changing smoothly and all is well.

The process of integration may now continue. The solution (16) of the second ballistic equation gives $\tan \theta_0 - \tan \theta_1$ (line 17). (In the example, this was done by a more elaborate formula than the plain arithmetic mean.) The result, subtracted from tan θ_0 (line 18), gives tan θ_1. The result is verified by computing the third difference (line 20).

The horizontal distance, $x_1 - x_0$, which the projectile covers in the interval $t_1 - t_0$ is found by a similar integration of the third ballistic equation and entered in line 21. The total horizontal range x_1 is found by subtracting $x_0 - x_1$ from x_0, entered in line 23 and checked for conformity with values in previous arcs by the third difference. The remainder of the column is computed in a similar fashion, and the final values for x_1, y_1, tan θ_1 and u_1 are transferred to the next column as x_0, y_0, tan θ_0 and u_0. The work then begins on the new arc.

This procedure is followed from the fourth arc to the end of the trajectory. The first three arcs are more difficult to compute and generally, considerable experience is required for the preliminary estimates, but there are certain empirical rules which make the task easier for the skilled computer. Many hundreds of trajectories have been calculated in this way, and the results show close agreement with firing trials whenever the resistance function has been accurately determined.

The reader will have noted several points of interest in the description given above. First, it is obvious that the actual computation does not call for any real knowledge of mathematics—the instructions are of the 'cookery book' type. The multiplications and divisions are usually done with a desk calculating-machine, and what is required of the computer is mainly accuracy in copying figures and a placid temperament—the latter is essential if it becomes evident, towards the end of the calculation, that a mistake has been made in one of the early arcs! Second, the various 'checks', or verifications, play an extremely important part, not only because estimation is an essential feature of the process, but also because of the risk of clerical errors. The best type of test is one in which the same quantity is calculated in two ways, using different ingredients, but the main purpose of the tests is to examine the assumptions for consistency.

Problems which lead inevitably to step-by-step solutions have little attraction for the pure mathematician—it was for this reason, no doubt, that Hardy once said that ballistics was not 'respectable'—but often they cannot be avoided, especially when, as in the case of air resistance, theoretical knowledge is lacking. The trajectory problem has been described at some length not only because it is an admirable example of the successful reduction of a difficult problem to elementary arithmetic, but also because it is a useful pointer to the future. The computer is provided with a *programme*, or set of instructions, and a *criterion*, that certain differences must be kept below pre-assigned limits. There is no need for him to know anything more, e.g. about the physical processes involved, and trajectory calculation could be performed perfectly by human beings who had never heard of gunnery—or even by *machines*.

Numerical Analysis and Calculating Machines

Computation has become now of such importance in advanced mathematics that a special branch of the science, called *numerical analysis* (not to be confused with *statistics*), is devoted to the subject. The aims and methods of numerical analysis differ considerably from those of ordinary analysis. In non-numerical mathematics a prime object is to expose the significant features of the pattern as a whole; numerical analysis is nearly always concerned with detail. The necessity for a special study arises because the complete analytical solution of a problem does not always supply a *practicable* answer for the engineer or physicist, or even for the compiler of mathematical tables.

Before considering these matters at greater length, it is necessary to interpolate here some remarks on the meaning which mathematicians attach to the words 'error' and 'mistake'. The difference is best explained by a simple example. The value of π correct to five places of decimals is

$$\pi = 3 \cdot 14159$$

The approximate value

$$\pi = 3 \cdot 142$$

involves an *error* not exceeding $+0 \cdot 0005$. In all numerical calculations involving π some error is unavoidable, because π, being an irrational number, cannot be expressed as a terminating decimal. In many calculations an error of $0 \cdot 0005$ in π is below the level of significance and the computation is correct within this margin. On the other hand, if in the course of a long calculation the computer transposed two digits and used the value 3·412 for π, this would be a *mistake* and the computation would be incorrect. (Transposition of digits in copying from one column to another or from a book of tables is a very common form of mistake, one which is especially difficult to locate in the subsequent search, for the same reason that mispellings of commonplace words are often unnoticed, even by professional proof-readers.)* Errors are inherent in numerical analysis; mistakes arise from human fallibility or faults in machines.

As a simple and well-known illustration of the gulf between ordinary and numerical mathematics consider the problem of calculating the natural logarithm of 2 (i.e. $\log_e 2$) correct to, say, three places of decimals. The solution provided by ordinary analysis is the series

$$\log_e (1 + x) = x - \frac{x^2}{2} + \frac{x^3}{3} - \frac{x^4}{4} + \cdots$$

and by putting $x = 1$ in this equation, we obtain a correct numerical formula for $\log_e 2$, namely

$$\log_e 2 = 1 - \tfrac{1}{2} + \tfrac{1}{3} - \tfrac{1}{4} + \cdots$$

However, to calculate $\log_e 2$ to three places of decimals from this series necessitates working out about 1000 terms! The numerical analyst gets over this difficulty by manipulating the original series to produce a difference formula

$$\tfrac{1}{2}\{\log_e (1 + x) - \log_e (1 - x)\} = x + \frac{x^3}{3} + \frac{x^5}{5} + \cdots$$

and then, putting $x = \tfrac{1}{3}$, we find another exact numerical expression for $\log_e 2$, namely

$$\tfrac{1}{2} \log_e 2 = \tfrac{1}{3} + \tfrac{1}{81} + \tfrac{1}{1215} + \cdots$$

This gives the value of $\log_e 2$ correct to the third place (0·693) with three terms only, and the saving in labour is obvious. The first step in any considerable computation is the arrangement of the expressions in a form adapted to computation.

Numerical analysis is not solely or even mainly concerned with the evaluation of formulæ, but also with problems for which a general analytical solution either cannot be found (as in the example of trajectory calculations) or, even if it can be

---

* The sceptical reader is assured that an example of this occurs in the paragraph he is reading.

found, is of no account in the specific problem. The main interest lies in devising a process to bring the problem within the range of known methods. Thus, in certain branches of applied mathematics it is necessary, very often, to find one root of an equation very accurately and the remaining roots very roughly or not at all. The equation may be algebraic (involving only powers of the unknown) or transcendental (involving functions such as sines, exponentials, etc.), and usually no general solution can be found. In this event the mathematician must first locate the required root by a graphical method, or by crude interpolation, and then refine this rough value by specialized techniques, such as iterative processes, of which the Newton–Raphson method is perhaps the best known.* At no stage in this work does an analytical expression emerge, only a sequence of numbers approaching a limit.

Numerical analysis also deals with problems of interpolation and extrapolation, approximate integration and differentiation, the solution of differential equations and other allied topics. All these can be handled by a computer using tables and desk calculating-machines, but in the present century there has emerged the modern calculating machine, of an automatic or semi-automatic type. This development amounts almost to a new conception of mathematical technique and is likely to prove a landmark in the history of science.

Mechanical devices to aid calculation are as old as mathematics. Children and uneducated people use their fingers in reckoning; for centuries, accounts were kept with the aid of notched sticks and the origin of the abacus or bead-frame is lost in antiquity. In European mathematics the *analogue* type of calculator has been studied extensively. An analogue machine is a mechanical or electrical device which uses a physical measurement to represent a number. The most common analogue calculator is the slide-rule, which is engraved so that intervals of length are proportional to logarithms, thereby reducing multiplication and division to addition and subtraction (that is, to the displacement of a length). Another convenient way of representing a magnitude is by rotation through an angle; this has the advantage that the rotating shaft can be made as long as we please, so that a magnitude can

* See Appendix VI.

be transferred from one place to another and easily transformed or operated upon by systems of gears. The polar planimeter, used for measuring areas, is an example of this type. The most elaborate analogue machine yet constructed, the modern *differential analyser*, consists essentially of planimeters (integrators) with provision for certain auxiliary operations, such as addition.

It is impossible, in the scope of this book, to give a detailed description of a differential analyser, and the reader must be referred to one of the many specialized texts now available for a full account.* A practical machine is necessarily very

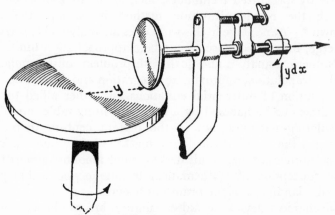

FIG. 19.   Wheel and Disc Integrator

large; that in use at the Royal Military College of Science occupies about 300 sq. ft. of floor-space and is driven by electric motors. The heart of the machine is the integrators, which are based on a mechanism invented by James Thomson (brother of Lord Kelvin) in 1876. Suppose it is desired to evaluate $\int y\,dx$, where y is some function of x which is known either by a formula or as a graph. The integrating device is shown diagrammatically in Fig. 19. A small vertical wheel, rigidly connected to a shaft, rests lightly on a rotating horizontal disc. If the wheel touches the disc at a point distant y units from the centre, rotation of the disc through a small angle dx turns the

* E.g. *The Differential Analyser* by J. Crank.   Longmans, Green & Co., 1947

wheel and its shaft by an amount proportional to $y\,dx$, provided that there is no slipping. If the wheel and shaft is moved horizontally according to the known variation of y while the disc is rotated steadily, the total rotation of the wheel is proportional to $\int y\,dx$, and this quantity may be conveyed by the integrator shaft to another part of the machine.

Before this device could be built into a practicable machine, a difficult mechanical problem had to be solved. The torque, or force causing rotation of the integrator shaft, arises entirely from friction between the wheel and the disc. This friction cannot be increased beyond a certain limit because the wheel must move easily backwards and forwards over the disc, and no slipping can be allowed. The solution lies in an ingenious adaptation of the capstan principle, known as a *torque amplifier*. The integrator shaft is connected by thin cord to a drum; the slightest increase in the tension causes the cord to be wrapped tightly around the drum, giving an enormous increase in tension at the output end. In this way the light touch of the wheel on the disc controls the whole power of the motor.

The main purpose of the analyser is to solve differential equations. For any given equation, the machine has to be 'set up' by making the appropriate connections and the data fed in by following a curve on the input tables. The result is produced either in the form of a curve drawn by the machine or, in more elaborate machines, as figures printed by an electric typewriter. In this way it is possible to solve in a few hours problems which would take a human being many days, but the time taken in setting up the machine can be quite long, and the advantage is evident only when many applications of the same equation are needed. The amount of error in the final result depends on the precision of the gear-cutting, absence of backlash and other mechanical details. All analogue calculators are limited by the accuracy with which the relevant physical quantities can be measured.

The *non-analogue* or *digital* type of automatic machine works on a different principle. We can illustrate this by considering first how a human being carries out a complicated calculation, such as the trajectory computation described above. The computer sits at a desk with a machine (slide rule, adding machine), paper to write on and mathematical tables to consult, and his brain to control the sequence of operations. He takes a number from the data of the problem, or from his work-sheets, associates this with a number drawn from the tables, performs certain operations (adding, subtracting, multiplying) and records the result on his work-sheet. We can regard the desk calculating-machine as an *arithmetical unit* which performs certain operations, the work-sheets and tables as a *store* and the man at the desk as a *control system* which selects the numbers and decides the sequence of operations.

The digital machine reproduces exactly these elements. It has an electrical arithmetical unit capable of performing the basic operations of addition, subtraction, multiplication and division at prodigous speed, and a store, or memory, which may take more than one physical form. The sequence of operations is decided by a *programme*, which has to be *coded* or translated into a form suitable for the machine. A calculation first of all has to be broken down into a series of relatively simple operations (such as addition) and the result related to the next operation. This is the programme. In the coding process instructions such as 'add the result of this operation to that from a previous operation and compare the result with another number' are represented by numbers, usually supplied on punched tape. The trajectory calculation described earlier in this chapter is an example of this process in a form well adapted for use by a human being; the process of integration is reduced to simple addition and division, and the instructions, although given in words, could be indicated, equally well, by numbers.

In the digital computer there is no measurement of a physical quantity and the accuracy attainable is limited only by the storage capacity of the machine. Naturally, such machines are extremely complicated, employing thousands of valves and demanding a specially trained staff of electronic engineers for their maintenance. The process of programming and coding calls for a unique combination of mathematical skill and knowledge of electronic devices, so much so that those who operate such machines represent, perhaps, the forerunners of a new type of scientist, the 'mathematical technician'.

4

AN ESSAY ON WAVES

Next, when I cast mine eyes and see
That brave vibration each way free;
O how that glittering taketh me!

HERRICK

Waves in the Natural World

IT IS A truism that a system, such as the physical universe, which exhibits both change and an enduring pattern of structure must also exhibit periodicity. Night and day, the ebb and flow of the tides, the phases of the moon and the alternation of the seasons mark the great stability of the solar system, and in the world of atoms and molecules, the universe which lies beyond the reach of our senses, known to us only by deduction, periodicity appears at every turn. In the nineteenth century the hard substantial atoms of early science dissolved into ætheric whirls and knots, and to-day the process has gone so far that only anonymous vibrations remain. The great change which has taken place in the present century in our conception of the ultimate constitution of matter is that the particle, as conceived by Newton and his immediate successors, has been replaced by the 'wave-packet', a kind of hybrid which shares many of the characteristic properties of particles and waves. We live, literally, in a universe of waves.

This change is reflected in the difficulty which confronts a physicist when he tries to define a 'wave' in modern usage. Intuitively, our ideas begin with waves on water (although these are by no means the simplest type), and it is only by ingenious experiments that we recognize similar characteristics in musical sounds, which are waves in air. The wave nature of light emerges only when even more sophisticated demonstrations are interpreted by mathematics, and the final stage, the

91

wave-like character of the elementary particles, such as electrons, brings us to the frontier where the concrete universe of the physicist and the abstract world of the mathematician seem to merge.

Perhaps the simplest way to indicate what the word 'wave' implies in modern science is to say that it is a *state of disturbance which is propagated from one place to another at a finite speed.* What passes is energy, in some recognizable form, and not matter (although in some types of waves there is a small permanent displacement of the medium). A stick floating on the surface of the sea is not carried forward permanently with a wave; what travels is the unevenness of the surface. When a gun is fired, the smoke remains near the muzzle, showing that the air itself does not travel, but the sound is heard later at great distances. A child, being asked to 'draw a wave', would produce what is usually called a 'wavy' line—mathematically a curve exhibiting regular oscillations about a mean position, like the graph of the sine function—and this may be taken as the basic representation of the simplest type of *waves of constant shape*, such as ordinary sound-waves and waves of light in a vacuum. On the other hand, the term 'shock wave', which has now become familiar because of the advent of high-speed air-craft, denotes a thin layer of air in which pressure and density suddenly jump to high values. This kind of wave is really a discontinuity which rushes along at a speed greater than that of sound (following an explosion or attached to a body moving at supersonic speed), and the nearest counterpart in water waves is the 'bore', or solitary wave of certain tidal estuaries, such as the Severn. Waves of this type, like water waves, do not preserve their shape as they move, and there is nothing obviously periodic in the phenomenon. Propagation, clearly, is the essential feature in the modern concept of a wave.

Mathematical Waves

To see what the mathematician has made of waves, we must begin with the simplest type. The sine function affords the basis for representing certain periodic motions, such as the (small) oscillations of a pendulum. We call this *simple harmonic motion.* Suppose we write

$$u = a \sin \frac{2\pi}{\lambda} (x - ct)$$

where λ and x are lengths, t is time and c is a velocity. A graph of u against x for a given t is the curve shown in Fig. 20 (this, of course, is simply a sine curve) and conforms to our intuitive idea of a wave.

There are certain technical terms in the study of waves which can be illustrated by this example. The distance between consecutive crests is clearly λ, because the sine function repeats its values every time the angle changes by 2π; λ is called the *wavelength*. The quantity λ/c is called the *period* of the wave, because the appearance of the wave relative to the origin is the same at times $t = 0$ and $t = \lambda/c$, and the crest

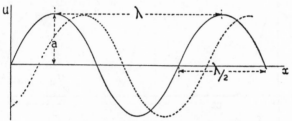

FIG. 20. Simple Harmonic Progressive Wave

moves the distance λ in the period. Instead of the wavelength the *frequency* n may be used to specify the wave—the relation is $\lambda n = c$.* If x is fixed and t is allowed to vary, the waves sweep past the point with the *velocity of propagation* c. The maximum value a is called the *amplitude*.

The curve of sines represents a *simple harmonic progressive wave*. We can generalize this without much difficulty, by noticing that the quantity $x - ct$ reproduces itself when t becomes $t + t'$ and x becomes $x + ct'$, for

$$x + ct' - c(t + t') = x - ct.$$

Hence any function of $x - ct$ can be said to represent a wave. This can be illustrated physically by a string lying on the axis of x.

---

* The reader is probably familiar with this method of indicating the position of radio stations on the dial of the receiver; here λ is usually given in metres and n in kilocycles per second, while c is the velocity of light.

If the string is displaced in any way perpendicular to the axis of x, we can write $u = f(x)$ for the shape of the resulting curve. If the displacements alter in such a way that the hump seems to travel with velocity c in the positive direction without change of shape, the equation representing the hump at any time t will still be $u = f(x)$, provided that we move the origin a distance ct in the positive direction (Fig. 21). The equation referred to the old origin will have x replaced by $x - ct$—i.e. $u = f(x - ct)$. This is the general equation of a wave of constant shape propagated in the positive direction with velocity c, and every wave of this type must be expressible in this form. Similarly, a wave going in the opposite direction (negative x) is represented by a function of $x + ct$.

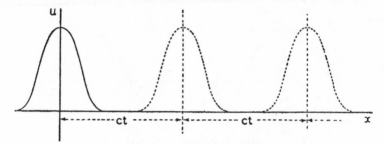

Fig. 21.   Illustrating $u = f(x - ct)$

These results have some interesting consequences. First, since any function of $x - ct$ can represent a wave, consider

$$u = e^{-(x - ct)^2}$$

where e is the base of natural logarithms. At $t = 0$, the outline is given by

$$u = e^{-x^2}$$

which is curve of the 'cocked hat' shape (well-known as the 'normal curve of errors'). If t is given a succession of values the whole curve moves along the x-axis without change of shape. This is a *solitary wave*, something like the shape of a rope lying on the ground and given a single up-and-down movement at one end (Fig. 22).

Second, consider the meeting of two simple harmonic progressive waves, identical in all respects but travelling in opposite directions. According to the result given above the disturbance of the medium is represented by

$$u = a \sin \frac{2\pi}{\lambda} (x - ct) + a \sin \frac{2\pi}{\lambda} (x + ct)$$

By ordinary trigonometry this is equal to

$$2a \sin \frac{2\pi x}{\lambda} \cos \frac{2\pi ct}{\lambda}$$

To see the physical implications of this expression, let u represent the elevation of the surface of a body of water. At a definite *point* (fixed x) the surface of the water, in general, moves up and down as $\cos (2\pi ct/\lambda)$. At a given *time* (fixed t) the form of the whole surface is a sine curve of amplitude

Fig. 22. The Solitary Wave

$2a \cos (2\pi ct/\lambda)$. The points for which $2\pi x/\lambda = n\pi$, where $n = \ldots - 2, - 1, 0, 1, 2 \ldots$ (i.e. points for which x is an integral multiple of half a wavelength) are always at rest in the mean surface—these are called *nodes*. Points of maximum displacement for a given value of t (i.e. where x is an odd multiple of the quarter wavelength) are called *antinodes* or *loops*. Such a disturbance is called a *standing* or *stationary wave*, but it should be noted that this form is not propagated, and therefore does not conform strictly with our previous definition of a wave. A standing wave is the result of combining two similar (but not necessarily identical) progressive waves travelling in opposite directions, and conversely, a progressive wave can be regarded as the combination of two systems of standing waves. Standing waves in water can be formed by tilting slightly a rectangular vessel and then restoring it to its original position. At each end of the vessel the water moves up and down the vertical sides, forming the antinodes.

Finally, it is easily shown by straightforward differentiation that

$$u = f_1(x - ct) + f_2(x + ct)$$

is a general solution of the equation

$$\frac{\partial^2 u}{\partial t^2} = c^2 \frac{\partial^2 u}{\partial x^2}$$

which is the equation of wave motion (Chapter 2). Here f_1 and f_2 are two arbitrary functions, and before further progress can be made it is necessary to give the initial and boundary conditions which specify a particular problem.

Two other matters must be discussed before we leave the kinematics of waves. The first is the question of *phase*. The character of the motion of the medium is determined by the amplitude and period, but we need also a zero of time if we are to specify the state of the motion at any instant. The quantity $\frac{2\pi}{\lambda} (x - ct)$ is called the *phase angle*, and we can choose any convenient time to fix its zero. (In physics we are nearly always concerned with phase *differences*, so that the choice of the zero of time can be arbitrary without causing inconvenience.) To say that two waves differ in phase means that we regard one as having started before the other, and the difference is sometimes given as a distance (e.g. a fraction of a wavelength) or as a time (e.g. a fraction of a period), as convenient.

The second matter is more subtle, and of the greatest importance in all studies of waves. So far we have considered only examples in which the wave velocity is independent of wavelength (or frequency). Systems in which the wave velocity is related to the wavelength are called *dispersive*, and in these there is a tendency for disturbances of different wavelength to travel in *groups* of approximately the same length. Groups of this kind are formed by waves on the surface of deep water, and by watching them it can be seen that the group, as a whole, does not travel as rapidly as the individual waves of which it is composed. The crest of a single wave can be observed to advance through the group and be lost in the smaller waves on approaching the leading edge.

To represent this mathematically, consider two trains of ·waves of the same amplitude but slightly different wavelengths.

For convenience, the quantity $2\pi/\lambda$ is replaced by the single symbol k, called the *wave number*, and the frequency n (reciprocal of period) is replaced by the *cyclic frequency* $\omega = 2\pi n$. Thus for simple harmonic waves

$$u = a \sin \frac{2\pi}{\lambda} (x - ct) = a \sin (kx - \omega t)$$

and clearly the wave velocity is simply ω/k. The disturbance caused by two such trains of wave numbers k, k' and frequencies ω, ω' is

$$u = a \{\sin (kx - \omega t) + \sin (k'x - \omega't)\}$$
$$= \left[2a \cos \left(\frac{k - k'}{2}x - \frac{\omega - \omega'}{2}t\right) \right] \sin \left(\frac{k + k'}{2}x - \frac{\omega + \omega'}{2}t\right)$$

If $k - k' = \delta k$, and $\omega - \omega' = \delta\omega$, say, this expression may be written

$$u = [2a \cos \tfrac{1}{2}(\delta kx - \delta\omega t)] \sin \{(k - \tfrac{1}{2}\delta k)x - (\omega - \tfrac{1}{2}\delta\omega)t\}$$

If δk and $\delta\omega$ are small, the sine term differs very little from $\sin (kx - \omega t)$, the form of the original wave, so that the expression represents something like a simple harmonic wave-train of variable amplitude. However, the sine term oscillates many times during one complete oscillation of the cosine terms, because for a fixed t, the amplitude $2a \cos \tfrac{1}{2}(\delta kx - \delta\omega t)$ varies very slowly with x. The complete picture at any instant looks like Fig. 23, oscillations rising to a maximum and then dying away. On the surface of the water there will be calm stretches separated by regions of disturbance. The individual waves obviously behave very much like the two original wave-trains and travel at much the same speed ω/k, but the groups behave very differently. The cosine term shows that the velocity of the group is $\delta\omega/\delta k$, which is approximately $d\omega/dk$. Since $\omega = 2\pi c/\lambda$, and $k = 2\pi/\lambda$, where c is the velocity of the original waves, it follows that

$$\text{group velocity} = \frac{d(c/\lambda)}{d(1/\lambda)} = c - \lambda\frac{dc}{d\lambda}$$

When c increases with λ, $dc/d\lambda$ is positive and the group velocity is less than the wave velocity. This is simply another way of saying that the constituent waves catch up with the group, grow in amplitude as they pass through it, and finally die away.   In a non-dispersive system c is not dependent on λ, $dc/d\lambda$ is zero and the group velocity and the wave velocity are the same.   The *length* of a group also is determined by the cosine term and is easily seen to be $\pi/(\tfrac{1}{2}\delta k) = 2\pi/\delta k$, so that

(length of group) × *(difference in wave number)* $= 2\pi$

For waves on the surface of deep water, the group velocity is approximately half the wave velocity, but as the water becomes more shallow, the two velocities coincide.   As we shall see later,

FIG. 23.   Wave Groups

the concept of group velocity is fundamental in *wave mechanics*, or the study of elementary particles by waves.

The phenomenon of wave-groups in a dispersive medium resembles in many ways another familiar effect of the super-position of vibrations, known as *beats*. If two sounds are produced which have much the same intensity (comparable amplitudes) but differ slightly in pitch, a throbbing is heard. (The effect is easily produced in the laboratory by two tuning-forks and is sometimes heard with twin-engined aircraft.) The compound wave is a vibration of slowly fluctuating amplitude, and the periodic rise and fall in intensity is heard as a distinct throbbing.   Sound-waves of small intensity in air are not dispersed, that is, the velocity of propagation does not depend on the pitch of the note.   If this were not so, the effect on the listener at the back of a large concert hall would be to make some music sound even stranger than it really is.

Fourier Series

So far we have discussed examples of the superposition of two wave-trains, for which the only mathematics required is a knowledge of the periodicity of the sine and cosine functions and the expression for the sum of two sines as a product of a sine and a cosine.   The reader will naturally ask what happens when more than two wave-trains are superimposed, and the answer is one of the most unexpected and revealing in the whole of mathematics.   In brief, it is that any continuous curve, however irregular, and many discontinuous curves, can be built up by adding together enough simple harmonic waves.

To show what this means, we begin by recalling to the reader the fact that the elementary functions of mathematics can be represented by *infinite series* of powers of the variable.   In advanced mathematics the power series is often chosen as the definition of the function.   Thus sin *x* is regarded, not as the ratio of the perpendicular and the hypotenuse (as in elementary trigonometry), but as the *name* of the series

$$x - \frac{x^3}{6} + \frac{x^5}{120} - \cdots .$$

It is easily shown that this series *converges*, that is, for any given value of *x*, the sum of the first *n* terms approaches a definite limit as *n* increases.   We can then say that sin *x* *exists*, that is, for any *x*, sin *x* has one definite numerical value. Another way of looking at it is to say that the sum of a finite number of terms of the series is an approximation to sin *x*; thus if *x* is small and measured in radians, sin *x* is approximately equal to *x*.   A better approximation for small *x* would be $x - \frac{x^3}{6}$, and so on.

Here we have a periodic function, sin *x*, approximated by powers of *x*, and the oscillation in sin *x* clearly is to be attributed to the alternation of sign in the series.   Can a non-periodic function, such as *x*, or any power of *x*, be approximated by a series of periodic functions?   Common sense says no, but common sense and intuition are notoriously unreliable in mathematics, and especially so here, for the answer is *yes*. We can put the problem in mathematical terms thus: suppose that a function $f(x)$ is known for all values of x between 0 and 2π. (The numbers 0 and 2π are introduced here purely for convenience; we can always arrange the notation so that 0 to 2π represents the range of x in question. Thus $f(x)$ might be the temperature of the air during a period of twelve hours, so that x is time, measured in units for which $\pi/6 = 1$ hour.) We want to approximate to $f(x)$ by a series of sines and cosines, that is, by the sum $s_n(x)$ where

$$s_n(x) = a_0 + a_1 \cos x + a_2 \cos 2x + \ldots + a_n \cos nx \atop + b_1 \sin x + b_2 \sin 2x + \ldots + b_n \sin nx \Bigg\} \quad \ldots(1)$$

The problem is to choose the coefficients $a_0, a_1, \ldots, a_n, b_1, \ldots, b_n$ so that the approximation is the 'best possible'. For this we have to give, in advance, some criterion of goodness of fit. The criterion chosen here is that known as *least squares*; if

$$\epsilon_n = f(x) - s_n(x)$$

we make the condition that the mean value of the square of ϵ_n in the interval $(0, 2\pi)$, or

$$\bar{\epsilon}_n = \frac{1}{2\pi}\int_0^{2\pi} \epsilon_n{}^2 dx = \frac{1}{2\pi}\int_0^{2\pi} \{f(x) - s_n(x)\}^2 dx$$

must be a minimum by the proper choice of the a's and b's.* This condition can be shown to imply that the coefficients are given by *Fourier's rules*,† viz.

$$a_0 = \frac{1}{2\pi}\int_0^{2\pi} f(x)\,dx$$

$$a_n = \frac{1}{\pi}\int_0^{2\pi} f(x) \cos nx\,dx \atop b_n = \frac{1}{\pi}\int_0^{2\pi} f(x) \sin nx\,dx \Bigg\} n = 1, 2, 3, \ldots$$

---

* This criterion may seem peculiar and even capricious at first sight, but a little reflection will show that it is very sound. If we laid down merely that ϵ_n should be small, the possibility of very large positive and negative differences would not be excluded, for these might cancel to produce a small total difference between $f(x)$ and $s_n(x)$. On the other hand, since $\epsilon_n{}^2$ is always positive, we can be sure that a small $\bar{\epsilon}_n$ always implies that the individual differences are small.

† This theorem is proved in many texts, e.g. Jeffreys, *Methods of Mathematical Physics*, Cambridge (1946), p. 419.

Knowing $f(x)$, either by a formula or as a series of values, the coefficients can be evaluated for all values of n. The reason for this particular form of the coefficients is that sin nx and cos nx are *orthogonal functions*, that is, they satisfy the following equations:

$$\left.\begin{array}{l} \int_0^{2\pi} \cos mx \cos nx dx = 0; \quad \int_0^{2\pi} \sin mx \sin nx dx = 0 \\[2mm] \int_0^{2\pi} \cos mx \sin nx dx = 0 \end{array}\right\} \text{if } m \neq n$$

and

$$\frac{1}{\pi}\int_0^{2\pi} \cos^2 nx dx = 1; \quad \frac{1}{\pi}\int_0^{2\pi} \sin^2 nx dx = 1.$$

There are many other examples of orthogonal functions to be found in higher mathematics, and the property is of fundamental importance in the theory of the differential equations of mathematical physics.*

So far all that has been stated is that if the coefficients are calculated according to certain rules, the trigonometric polynomial (1) will furnish a good *approximation* to $f(x)$. Can the function ever be *perfectly* represented by a series of sines and cosines, for example, by letting the number of terms become infinite? If $f(x)$ is continuous and of bounded variation, the answer is given by *Fourier's theorem*: the infinite series

$$a_0 + a_1 \cos x + b_1 \sin x + a_2 \cos 2x + b_2 \sin 2x + \ldots \ldots (2)$$

converges to the function $f(x)$ at all points whenever the coefficients satisfy Fourier's rules. The series is then called the *Fourier expansion of $f(x)$* and the a's and b's are called the *Fourier coefficients*. Let us see what this means by looking at a simple example.

Fig. 24 shows what is sometimes called a 'triangular wave', which we shall try to represent by a Fourier series. This is a continuous function which decreases linearly as x changes from 0 to a, then increases linearly between a and $2a$ and so on. (Despite the sharp changes in direction at $x = 0$, a, $2a$, ..., the function is continuous because as we approach any point

---

* The reason for Fourier's rules is fairly obvious once the orthogonal property is pointed out; all that one has to do is to multiply $f(x)$ by cos nx and sin nx in turn, do the same to the terms in the polynomial and integrate.

from either side we reach the same value.*)   The period of the function is $2a$; we can change this into 2π by writing ξ for $\pi x/a$—this is a trivial change which does not complicate the calculations in any way.   First, we note that the function is *even*, that is, $f(-x) = f(x)$.   Now cos $(-x) = \cos x$, but sin $(-x) = -\sin x$, so that even functions are represented by series of cosines and odd functions by series of sines.   (Most functions are neither odd nor even, but any function can be represented as the sum of an odd and an even function.)   To calculate the coefficients a_n we put $f(x) = K(a-x)$ in the

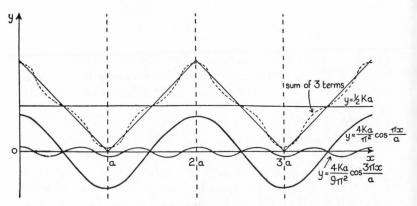

FIG. 24.   The Triangular Wave and its Representation by Harmonic Waves

interval $(0, a)$ and $f(x) = K(x-a)$ in the interval $(a, 2a)$, change the variable to make the interval $(0, 2\pi)$ and integrate according to Fourier's rules.   Thus

$$a_n = \frac{Ka}{\pi}\int_0^\pi \left(1 - \frac{\xi}{\pi}\right) \cos n\xi \, d\xi + \frac{Ka}{\pi}\int_\pi^{2\pi} \left(\frac{\xi}{\pi} - 1\right) \cos n\xi \, d\xi$$

$$= \frac{2Ka}{n^2\pi^2}(1 - \cos n\pi)$$

Now cos $n\pi = 1$ if n is even and is -1 if n is odd.   Hence

$$a_n = \frac{4Ka}{n^2\pi^2} \text{ if } n \text{ is odd}$$

$$= 0 \text{ if } n \text{ is even.}$$

* See Appendix II.

Thus the Fourier series is

$$\frac{1}{2}Ka + \frac{4Ka}{\pi^2}\left\{\cos \pi \left(\frac{x}{a}\right) + \frac{1}{9}\cos 3\pi \left(\frac{x}{a}\right) + \frac{1}{25}\cos 5\pi \left(\frac{x}{a}\right) + \dots\right\}$$

and because $f(x)$ is continuous and of bounded variation, Fourier's theorem assures us that the series accurately represents the 'triangular wave'. Let us see how this happens, by looking at the successive approximations. The first term is $\frac{1}{2}Ka$, which is simply the average value of $f(x)$ throughout the whole range, and is a very rough approximation to $f(x)$. The second term, $\frac{4Ka}{\pi^2}\cos \pi \left(\frac{x}{a}\right)$, is a simple harmonic wave of period equal to the period of the function. When this is added to $\frac{1}{2}Ka$ to produce the second approximation, the wave is lifted above the x-axis by the amount $\frac{1}{2}Ka$, and the result is a much better approximation. The third term, $\frac{4Ka}{9\pi^2}\cos 3\pi \left(\frac{x}{a}\right)$, is another simple harmonic wave, but of much smaller amplitude and of period $\frac{1}{3}$ of that of the function. The sum of the first three terms gives a very good approximation to the triangular wave, and as more terms are included, it becomes increasingly difficult for the draughtsman to distinguish between the straight and wavy lines.

The series (2) also can be written in the form

$$f(x) = a_0 + A_1 \cos (x + \phi_1) + A_2 \cos (2x + \phi_2) + \dots$$

(or as a series of sines), where

$$A_n = \sqrt{(a_n^2 + b_n^2)}$$
$$\tan \phi_n = - b_n/a_n$$

Fourier's theorem shows that any continuous periodic function is built up of an infinite number of waves (called *harmonics*), of integral multiples of a fundamental frequency and differing phase. Fourier's rules for calculating the coefficients are really rules for arranging the amplitudes and phase differences so that the sum of all the waves (that is, the result of super-position of harmonics) gives exactly the desired shape. Each successive harmonic removes a little more of the crudeness of the earlier approximations. In the example given above, the function $y = K(a - x)$ has been made periodic by continuing it beyond $x = a$, but essentially what is done is to represent a non-periodic function in a *finite* interval by waves. Any continuous curve can be represented in this way, although naturally a very irregular curve needs many terms for a really close fit.   To take an example from meteorology, the temperature of the air near the ground during 24 hours of fine weather reaches a maximum in the early afternoon and a minimum just before dawn.   Such a variation is very poorly represented

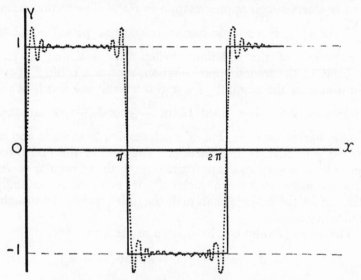

Fig. 25.   The Square Wave and Gibbs' Phenomenon

by a single mean value; the sum of the mean value and the first harmonic is better, but for accurate work three or even four harmonics are required.   The computation of the Fourier coefficients is done either directly or by machines called *harmonic analysers*.

Let us now consider the important type of discontinuous function shown in Fig. 25.   The discontinuity arises because on approaching, say, $x = 0$ from the left we tend to the value -1, but an approach from the right gives $+1$.   Such points are called *finite discontinuities*.   This function, called the 'square wave', is of importance in radio and other branches of engineering—it gives an idealized picture of a series of impulses, like switching. Fourier's rules give the series

$$\frac{4}{\pi}\left(\sin x + \frac{1}{3}\sin 3x + \frac{1}{5}\sin 5x + \ldots\right)$$

the constant term (a_0), which is the average value of the function, being zero. The earlier approximations behave very much as we are led to expect from the example of the continuous function, but there is a peculiar feature at the points of discontinuity. As the higher harmonics are added to the approximation, the flat portion of the wave is reproduced in the usual manner, but at the points of discontinuity there is a permanent overshoot. The main effect of adding yet more terms is to narrow the loop of the overshoot until it becomes a 'spike', which is never shorter than about nine per cent of the magnitude of the jump of the function at the discontinuity. This feature is known as *Gibbs' phenomenon*, after the American physical chemist Willard Gibbs, who first described the anomalous behaviour of the approximation in a letter to *Nature* in 1899. Gibbs' phenomenon is important in radio theory in relation to the problem of 'cut-off'.

The convergence of a Fourier series poses many difficult and delicate problems. Although we have discovered much about the behaviour of these series we still do not know, for example, the necessary and sufficient conditions to ensure convergence. When a modern pure mathematician speaks of the 'Fourier series of a function', he means that the coefficients are formed from the function by Fourier's rules, and nothing more. He does not imply that the function 'equals' the series in any sense, although in many cases the series does converge to the function. The sum of a Fourier series at a point x_0 is the limiting value of the sum of the first n terms, with x replaced by x_0, as n becomes indefinitely large. This limit may or may not be equal to $f(x_0)$. Again, not every trigonometrical series is a Fourier series, and even if a series is a Fourier series (that is, one with coefficients determined by Fourier's rules), it need not be the Fourier series of its sum. The reader will appreciate that the general problem of Fourier series is as subtle as any in mathematics.

We have seen how to decompose a function into its waves in

any *finite* interval. What happens when the interval is *infinite?* Without entering into details, it can be said that in this case the series becomes an integral. The *Fourier integral representation* of $f(x)$ is

$$f(x) = \int_0^\infty a(n) \cos nxdn + \int_0^\infty b(n) \sin nxdn$$

with the quantities $a(n)$ and $b(n)$ given by

$$a(n) = \frac{1}{\pi} \int_{-\infty}^\infty f(x) \cos nxdx; \quad b(n) = \frac{1}{\pi} \int_{-\infty}^\infty f(x) \sin nxdx$$

These equations are equivalent to

$$f(x) = \int_0^\infty A(n) \cos [nx + \phi(n)]dn$$

where

$$A(n) = \sqrt{\{a^2(n) + b^2(n)\}}$$
$$\tan \phi = - b(n)/a(n)$$

as in the series.

The Fourier integral representation involves all frequencies, because the integration with respect to n is from zero to infinity. (In the series representation only those frequencies which are multiples of the fundamental frequency occur, like a note and its overtones.) Thus we can say that a non-periodic function also has a frequency composition.

Radio Signals and their Reproduction

The function $A(n)$, introduced above, corresponds to the A_n of the series, and a graph of $A(n)$ against n shows the composition of the function, in terms of frequency, just as the spectroscope reveals the composition of white light. We call graph of $A(n)$ against n the (amplitude) *frequency spectrum* of the function. Let us see what this means by considering some examples from radio.

A modern radio set is designed to receive signals over a certain band of frequencies on either side of the fundamental frequency to which it is tuned, and to reject all others. An ideal receiver, when tuned to a frequency n_0, reproduces faithfully all signals composed of frequencies between

$n_0 - m$ and $n_0 + m$, and the difference $2m$ is called the *band-width* of the receiver. The problem we shall now investigate is ultimately that of finding the correct band-width for a given signal, so that it may be reproduced without excessive distortion.

If the signal consists of an unending train of simple harmonic waves of the same frequency, the receiver need only respond to this frequency and the band-width required is zero. So much is obvious, but if we go to the other extreme and consider a signal consisting of a single wave $f(t) = \cos n_0 t$ transmitted over a short interval t_1 to t_2, and zero elsewhere, we find a somewhat unexpected result. To ascertain the frequency spectrum of this signal we must evaluate the integrals for $a(n)$ and $b(n)$ given above, remembering that the amplitude of the wave is zero up to time t_1, is unity from t_1 to t_2, and zero again from t_2 to infinity. Thus

$$a(n) = \frac{1}{\pi} \int_{-\infty}^{\infty} f(t) \cos nt\, dt = \frac{1}{\pi} \int_{t_1}^{t_2} \cos n_0 t \cos nt\, dt$$

and similarly for $b(n)$. In this way we can calculate $A(n) = \sqrt{\{a^2(n) + b^2(n)\}}$ and plot the result as a function of n for any values of $t_2 - t_1$. The result shows the fraction of the signal amplitude which is made up of waves of any given frequency. If $t_2 - t_1$ is so short that only one loop of the wave is included, the graph of $A(n)$ is a broad-based curve extending over a wide range of frequencies on either side of n_0 (Fig. 26 (a)). As we lengthen $t_2 - t_1$, so as to include more and more cycles of the function $\cos n_0 t$, the frequency curve becomes more tightly packed about n_0 (Fig. 26 (b)), and finally, for an infinite train, the curve shrinks to a single line at n_0. The required band-width is the range of frequencies which takes in as much of the curve as the set designer considers necessary to secure good reproduction. Fig. 26 shows frequency spectra for a single harmonic loop, a short train, a single rectangular pulse and a long train.

Thus to reproduce faithfully a short pulse which starts very suddenly (theoretically instantaneously) and finishes equally abruptly requires a receiver of very great band-width, because it is clear from diagram (c) that the higher frequencies are present in substantial amount, and to exclude them would mean losing much of the detail of the original curve, especially at the corners. This is simply another way of saying that to get a good approximation to a 'difficult' curve (and difficult here means chiefly sharp corners) it is necessary to use very many constituent waves. If the signal is periodic, like a succession of rectangular pulses (e.g. a square wave), the

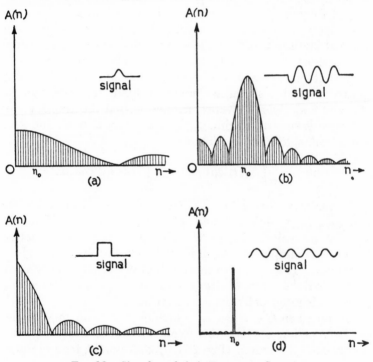

Fig. 26. Signals and their Frequency Spectra

frequency spectrum is obtained from the series in a similar manner, and shows similar characteristics.

These conclusions have important consequences in television and radar, where the signal is usually a sequence of sharply cut-off pulses. If the band-width of the television receiver is too small, much of the detail of the picture will be lost, and in radar the distortion may seriously affect the accuracy of range-finding. (In ordinary sound broadcasting the corresponding problem is that of reproducing transients, such as the 'bite' of the bow on the violin string.) To secure a large band-width without bringing in interference necessitates the use of very high frequencies, and it is for this reason that television stations transmit on short waves.

We can carry this investigation further by introducing a new version of the Fourier integral representation, one which is rapidly becoming a favourite tool of the applied mathematician. We begin by reverting to the algebra of complex numbers (Chapter 2), starting with one of the most famous identities in mathematics,

$$e^{i\theta} = \cos \theta + i \sin \theta$$

where e is the base of natural logarithms and i, as usual, is $\sqrt{-1}$. Let $F(\nu)$, where $\nu = n/2\pi$, be defined as a kind of 'vector' version of $A(n)$ by

$$F(\nu) = \pi\{a(n) - ib(n)\}$$

so that the spectrum function $A(n)$ is simply the modulus, or absolute value, of $F(\nu)/\pi$ (p. 31). It can be shown without difficulty that the Fourier integral representation of $f(t)$ is equivalent to the *reciprocal relations*

$$f(t) = \int_{-\infty}^{\infty} e^{2\pi i\nu t} F(\nu)\,d\nu$$

and

$$F(\nu) = \int_{-\infty}^{\infty} e^{-2\pi i\nu t} f(t)\,dt$$

The functions f and F are then said to be *Fourier transforms* of each other. Thus the Fourier transform of a function is another way of expressing the frequency spectrum of the function, and often furnishes the quickest method of finding the frequency composition of a signal. There are also other uses for, and other kinds of, integral transforms, and the theory plays a leading part in the modern treatment of differential equations, so much so that it has been worth while compiling dictionaries of transforms. From these, given f, the transform F can be written down without the labour of calculation.

Among the various properties of transforms there is a general principle which points the way to one of the most widely discussed laws of atomic physics. It is a general rule that if the graph of one member of a Fourier transform is a compact, sharply cut-off curve, the other is widely spread. Thus, in the case of a long train of sinusoidal waves the frequencies are closely clustered around the fundamental, but as the train shortens in time, the spread of frequencies increases. This leads to an important rule for the design of radar and other pulse-operated systems, that for good reproduction the band-width of the receiver must be kept in inverse proportionality to the duration of the pulse. For accurate range-finding by radio echoes, two requirements must be met. First, the pulse must have a steeply rising leading edge, so that the instant at which transmission begins can be easily identified. This is necessary to ensure *accuracy*. Second, to distinguish the target from other reflecting objects at nearly the same range, the pulse must be very short, otherwise one echo may be completely masked by the other. This is simply another way of saying that the *resolving power* of the set must be as high as possible. The length in space of a pulse is the duration multiplied by the velocity of light and, combining the two requirements, steepness of shape for accuracy and narrowness for good resolution, it follows that a practicable radar set must generate very brief pulses, usually of the order of fractions of a micro-second. This requirement, in turn, means that the receiver must be designed to accept a wide range of frequencies. If the duration of the pulse is Δt and the width of the main part of the $(A(n), n)$ curve is Δn, the rule is

$$\Delta t \, . \, \Delta n \simeq 2\pi$$

We can put the matter in another way, which is useful when we consider the physics of the atomic world. We suppose that in the course of an experimental investigation there is need for a very accurate time signal, to be transmitted as a radio impulse of definite wavelength. The two requirements are, strictly, incompatible, and this is a fundamental property of waves, not to be evaded by ingenuity in the design of apparatus. If the signal is to be transmitted at a definite time, it must be a sharply shaped pulse of very brief duration, and this means that its frequency composition is very broad. On the other hand, if we make the frequency precise by sending a long train of simple harmonic waves, we must be vague about time.

The time element may be regarded also as specifying position, so that position and wavelength cannot be defined with unlimited accuracy at the same time, and a wave-group is subject to a fundamental indeterminacy. There are many examples of this kind of reciprocity to be found in physics, where any attempt to enhance one feature causes a corresponding reduction in an associated feature—increase of magnification in a telescope causes loss of field, and so on. In plain language the result is expressed in the old proverb that what is gained on the roundabouts is lost on the swings. Nature never gives something for nothing. We shall see later that this principle, when applied in atomic physics, brings us face to face with what seems to be one of the ultimate limits of human knowledge.

Waves and Particles

In 1900 the German scientist Max Planck published an account of some conclusions he had reached concerning the nature of radiant heat, and in doing so turned the main stream of physics. In the previous century the conception of light, and of radiation generally, as a continuous stream of vibrations seemingly had carried all before it, except for a partial failure to account for the distribution of energy in the radiation from hot bodies. Planck showed that in this problem theory and experiment could be reconciled by adopting the hypothesis that exchanges of energy between the body and the radiation take place in an intermittent fashion. He visualized energy being emitted or absorbed in minute packets (*quanta*) of amount hn, where n is the frequency of the tiny oscillators which make up the radiating surface and h is a new physical constant, with a value about 6×10^{-27} erg sec. This means that energy is atomic in structure. The quantity h, now known as *Planck's constant of action*, is fundamental in the *quantum theory*.

The atomic nature of electricity had been demonstrated in 1897, when J. J. Thomson deflected cathode rays* by an electrostatic field. The atom of electricity is called the *electron*, and Thomson was able to assign definite values to the ratio of its mass and charge, and even to estimate its radius.

* Cathode rays are produced when a high voltage current of electricity is passed through an evacuated tube.

When a stream of electrons strikes the anticathode in a highly evacuated bulb, X-rays are produced, and it is well known that these rays behave like a beam of light. (It is this property which makes X-rays useful in medicine.) When X-rays, in turn, are allowed to strike negatively charged metal they cause a loss of charge by liberating electrons, and it is found that the energy of the electrons so emitted is substantially the same as that of the electrons which produced the X-rays. No energy is lost in the process of going from electron to X-ray and back again, so that it is difficult to resist the conclusion (first stated by Einstein) that the energy in X-rays also is carried in little packets, and that these packets do not spread, like waves. Such packets of energy, atoms of light (using the word 'light' in a very general sense) are called *photons*, and the energy E associated with a photon also must be hn. We can also associate momentum (p) with a photon, for in the particle theory, momentum is energy divided by velocity. Clearly, the only velocity we can associate with the photon is that of light (or any other electromagnetic radiation), c, so that

$$p = \frac{E}{c} = \frac{hn}{c} = \frac{h}{\lambda}$$

since c/n is the wavelength λ. We have thus found a possible expression for the momentum of an atom of radiation in terms of the wavelength of the radiation and Planck's constant. Finally, in 1922 A. H. Compton discovered that when a photon meets an electron, the energy is redistributed partly in a recoil and partly in scattered radiation of lower frequency. The Compton effect, as it is called, completed the picture by showing that momentum also is atomic in structure, so that waves and particles seem to have become inextricably mixed. The problem which agitated physics in the early years of this century was that of reconciling two such fundamentally different conceptions as the particle and the wave.

When faced with a dilemma of this kind, it is as well to go back to the foundations. Why do we believe that light is made up of waves? In a beam of light there are no visible crests and troughs to indicate vibrations, as in water. We can measure the relative intensity of light very simply, and we can measure something we call the velocity of light, but such quantities could be associated quite as easily with a shower of particles. Intensity, for example, could be interpreted as equivalent to the number of particles crossing a unit area in the path of the beam in unit time, just as we do with a jet of smoke. The real reason why we favour the wave hypothesis for light is that a wave is the ideal mathematical model for representing what happens when two rays of light cross, or when light is made to form certain patterns by being reflected from a sheet of polished metal which has been ruled with very fine lines, set close together (i.e. a diffraction grating). It is then found that the wave hypothesis predicts the patterns perfectly, and our belief in the vibratory nature of light is strengthened when we find that we can produce similar interference patterns by undoubted ripples in a tank of liquid. These measurements also yield consistent values for what we choose to call the wavelength (or frequency) of the light. Thus the evidence for the wave nature of light turns out to be no more than the statement that the mathematical theory of waves gives correct predictions when applied to visible light or any other form of radiation.

The same is true of particles. We saw in Chapter 2 that the 'fluid particle' is a figment of the mathematician's imagination, and so is any other particle in the Newtonian sense of the word. There are problems for which the particle is the appropriate basis, and for which it would be foolish to use any other model. Thus although we could represent, if we wished, the trajectory of a cannon-ball by a Fourier series or integral, that is, by waves, it would be a waste of time to do so, because the problem is solved much more easily on the particle basis. There are no 'wave mechanics' in external ballistics.

The mathematician regards particles and waves simply as means to an end, and he is unlikely to be worried by the impossibility of forming a mental image of something which is simultaneously a wave and a particle. To him particles and waves are as real as points in Euclidean space or the operator i, no more and no less. His concern is mainly that the chosen method of representation gives correct predictions. In 1924 the French mathematical physicist Louis de Broglie proposed that any moving elementary entity, such as an electron, which is known to have the properties of a particle (e.g. momentum) should be associated with a train of waves, the connection between the two systems being the relation established for photons

$$p = \frac{h}{\lambda} \quad \text{or} \quad \lambda = \frac{h}{p}$$

De Broglie's suggestion to the physicists was therefore that if a beam of electrons were allowed to strike a suitable grating, diffraction effects would be observed just as if the beam were a light-ray of wavelength $\lambda = h/p$, where p is the momentum of the particle. The difficulty in carrying out this experiment lies mainly in finding a suitable grating, for the distance between the lines ruled on the grating must decrease with decreasing wavelength, and de Broglie's waves could only be of very short length. (To fix ideas, consider the de Broglie wavelength of a 60 kV electron beam. This is about 0·05 Å, whereas visible light has wavelengths of the order of 4000 Å.) Nature, however, has provided a perfect grating, far finer than can ever be produced by a ruling engine, in the crystal, which is a three-dimensional lattice of atoms. In 1927 Davisson and Germer bombarded a crystal of nickel by a beam of slow electrons and found all the diffraction effects which de Broglie predicted. When ordinary light is reflected from a grating, maxima and minima of intensity are observed, and it is a familiar laboratory exercise to calculate the wavelength of the light from the measured distances between the maxima. The wavelengths of the electronic waves, calculated in the same way, agree with the equation $\lambda = h/p$. Since 1927 there have been many other verifications of de Broglie's suggestion, even for molecules of hydrogen and helium when diffracted by a crystal. The most striking example of the wave character of the electron, is afforded, of course, by the electron microscope.

We can make one further interesting deduction from de Broglie's correlation of particles and waves. The expression for group velocity (p. 97) may be written as $-\lambda^2 \frac{dn}{d\lambda}$, and if we put $E = hn$ and $\lambda = h/p$, this expression becomes simply dE/dp. In the classical mechanics of particles

$$p = mv; \quad E = \tfrac{1}{2}mv^2 + \text{(potential energy)}$$

where v is the speed of the particle. Comparing the two expressions we get at once the result that *the velocity of the particles in a beam of atoms or electrons is equal to the group velocity of the associated de Broglie waves.*

It is implied above that the mathematician adopts a somewhat detached attitude in his choice of particles or waves to represent physical phenomena, but this is over-simplifying matters. Getting the right answer on one occasion is not enough—as any schoolboy knows—and in associating waves with particles we must be prepared to accept all the consequences of such a step. We may not select certain wave properties and ignore others as inconvenient without very careful consideration, for our aim is to produce a logical system which can be trusted at all points. There is a temptation to say that a group of waves can represent a particle and to leave it at that, but there are certain inherent properties of short wave-trains which appear at first sight to be incompatible with the particle concept. Yet it turns out that one of these properties is exactly what is needed.

The Newtonian particle has a definite position at a definite time; we may therefore say that it has precise values for momentum and position at all times. Are these properties appropriate to elementary bodies such as electrons? We can find the position or the momentum of an elementary particle only by means of light, either emitted or reflected. But light-pulses cannot be formed more rapidly than about a hundred million times a second, and we could conceive particles oscillating even faster than this. The particle would then be like a ship at sea which can report its position by radio only at certain hours, the intervening time being taken up in recharging its batteries. In between signals all that the observers on shore can do is to draw a circle around the last reported point and say that the ship is somewhere in the circle. In other words, their statements are couched in the language of *probability*, and the same must be true of atomic physics. There is fogging in the picture of an electron which belongs to nature, and any mathematical system adopted to describe electrons should include such fogging as an intrinsic property.

These considerations rule out the Newtonian particle as the basis of a mathematical system for representing electrons and other elementary entities—it goes too far and attributes a precision to our knowledge which is not justified. The wave-picture, on the other hand, has the quality of 'imperfect specification' built into it, and this is one of the reasons why it succeeds so well in atomic physics. We have seen that a radio signal cannot be perfectly precise in time (or what is much the same thing, position) and in frequency at the same time, and if we become more precise about one, we become increasingly vague about the other. In the de Broglie scheme frequency, being proportional to $1/\lambda$, is equivalent to momentum, and the property we have discussed for radio waves becomes for elementary entities the 'principle of uncertainty' (*Unbestimmtheitsprinzip*) of Heisenberg, which has worried so many philosophers. Heisenberg's principle is this: if we measure simultaneously the position (x) and the momentum (p) of a particle we shall have certain errors Δx and Δp. If the results of these measurements are described by a function having the form of a wave-group, it follows that the product $\Delta x \Delta p$ is of the order of h, Planck's constant. No measurements can be made for which $\Delta x \Delta p$ is less than h. Thus in pledging ourselves to a belief in the validity of the mathematical wave as a means of representing measurements of elementary entities we are obliged to accept Heisenberg's principle, and conversely. This is the foundation of *wave mechanics*. The question 'Is an electron a particle or a wave?' has about as much meaning as one asking if an electron is a quartet of numbers or a sine. An electron is a physical entity for which wave theory forms a useful calculus, and the possibility of some other calculus being even more useful in the future cannot be ruled out, so that there is no finality to our picture.

We can put the matter in another way, which in some respects is more illuminating. The instruments of the physical laboratory are the receivers with which we pick up signals from the atomic world. We cannot yet read all the signals, but it appears that the language used is that of mathematics, and the dialect which is most favoured is that of wave theory. Interpreted in this way the messages make sense, but we are not able to say much about the entity which produces the signals, except that we can predict, in many instances, the kind of signal which will be evoked by a suitable stimulus. In other words, we are beginning to know the characteristics of the broadcaster even if we cannot imagine what he really looks like.

Eigenfunctions and Eigenvalues

Let us turn now to music, an appropriate choice, because it was the realization of the connection between musical harmonies and whole numbers which led Pythagoras to regard mathematics as the key to the riddle of the universe. One of the most ancient and simple of musical instruments is the stretched string, which may be struck, bowed or plucked, and yet in the investigation of this simple mechanical system we find once more links with the most abstruse conceptions of modern atomic physics and (as we shall see later) even with high-speed flight.

We begin with a string kept under tension by being fixed at two points, $x = 0$ and $x = l$. At time $t = 0$ the string is displaced at the point x by a small amount u in the direction perpendicular to the axis of x. When released, the string flies back with an acceleration $\partial^2 u/\partial t^2$. The disturbance gives rise to a force trying to restore the system to its average value ($u = 0$) and by the general theorem given on p. 66, $\partial^2 u/\partial t^2$ must be proportional to the Laplacian of u, which in this case is simply $\partial^2 u/\partial x^2$. The factor of proportionality turns out to be the tension divided by the density of the string; this has the dimensions of velocity squared and is written c^2 for convenience. The equation of motion of the string is thus

$$\frac{\partial^2 u}{\partial t^2} = c^2 \frac{\partial^2 u}{\partial x^2} \qquad \ldots(3)$$

which will be recognized as the simplest form of the wave-motion equation (p. 65). Thus waves must appear when a taut string is displaced from its rest position and released.

To solve eq. (3) the mathematician employs a process not unlike that used by chemists when analysing a compound. This is known as *separation of variables*. It is assumed, as a trial hypothesis, that the solution can be expressed as the product of a function of x alone, say $X(x)$, and of t alone, say $T(t)$. After substitution of $u = X(x) T(t)$ the partial differential equation (3) becomes

$$\frac{1}{c^2 T} \frac{d^2 T}{dt^2} = \frac{1}{X} \frac{d^2 X}{dx^2}$$

The left-hand side of this equation is a function of t only, and the right-hand side depends only on x; further, the equation must hold for all values of x and t. The only possibility is that both sides are equal to the same constant. This reduces the equation to two ordinary differential equations, both very well known,

$$\frac{1}{X}\frac{d^2X}{dx^2} = \text{constant} = -k^2, \quad \text{say} \qquad \ldots(4)$$

and

$$\frac{1}{T}\frac{d^2T}{dt^2} = -k^2c^2 \qquad \ldots(5)$$

Eq. (4) can be solved, without difficulty, in the form

$$X = A \sin(kx + \epsilon)$$

where A and ϵ are arbitrary constants. The physical conditions of the present problem are

$$u = 0 \quad \text{for} \quad x = 0 \quad \text{and for} \quad x = l$$

which say simply that the string is fastened at both ends. To satisfy these conditions we could, of course, put $A = 0$, but this would not help because u would then be zero everywhere and always, and we are left with ϵ, which clearly must be zero if u is to vanish when $x = 0$. But now $u = A \sin kx$ will not vanish for $x = l$ unless we impose some condition on k, and $\sin kl$ will vanish if, and only if, k takes the values 0, π/l, $2\pi/l$, ..., $n\pi/l$, where n is a whole number. The value $k = 0$ must be rejected because it leads to the trivial result that $u = 0$ everywhere, and we conclude finally that fastening the string at both ends means that the x-part of the solution must have the form $\sin(n\pi x/l)$. Having fixed k, the time part of the solution must involve $\sin(n\pi ct/l)$ and $\cos(n\pi ct/l)$.

We can now identify the solution physically. Every possible value of k, that is, $k = n\pi/l$ where $n = 1, 2, 3, \ldots$, gives rise to a wave of length $\lambda = 2\pi/k = 2l/n$, so that we have systems of standing waves on the string. The fundamental mode has nodes only at the end points ($\lambda = 2l$); the first harmonic has a node at the mid point of the string ($\lambda = l$) and so on. In other words, we have discovered by mathematical analysis what is well known to every student of physics, that a taut string vibrates in a well-defined system of standing waves.

The wave numbers k can thus take only certain values characteristic of the system. These are called *eigenvalues* (the hybrid of German and English is used because the adjective 'characteristic' tends to be overworked in mathematics), and the characteristic waves sin $(n\pi x/l)$ are called *eigenfunctions*. These concepts are so important in mathematical physics that it is worth examining the solution a little more deeply. The outstanding fact about a vibrating string is that not every frequency (or wavelength) appears, otherwise a plucked string would give out, not a musical note, but a *noise*, that is, a jumble of waves of all lengths. Waves of all kinds are hidden in the

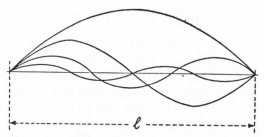

FIG. 27. Standing Waves (Fundamental and Overtones) on a String

differential equation, and the boundary conditions pick out those which conform to physical reality in a given situation. There are few examples which show so well the power of mathematical analysis; the use of stringed instruments for music depends on the fact that the eigenfunctions of this problem consist of a fundamental note and its overtones.

So far we have dealt only with the space variation. The time variation, as we have seen, involves $n\pi ct/l$ as the argument of the sine and cosine terms. The frequency of this oscillation is $n\pi c/l$, which increases as c is increased (i.e. as the tension increases) or as l is reduced. Thus tightening the string or reducing its length (stopping) raises the pitch of the note, and this can be done in steps of any size. Hence we can make the fundamental note have any pitch, but each note will be accompanied by its overtones, which partly determine the quality of the sound.

The problem is completely solved when we know how to calculate the subsequent motion when the string is given *any* initial displacement and released from rest. These are the *initial conditions*. At $t = 0$ let the shape of the string be defined by $u = f(x)$, where $f(x)$ is any function which is zero at $x = 0$ and at $x = l$. The general solution is the sum of all the particular solutions, or

$$u = \Sigma \sin (n\pi x/l)\{a_n \cos (n\pi ct/l) + b_n \sin (n\pi ct/l)\}$$

where a_n and b_n are numbers to be determined. The string starts from rest, therefore $b_n = 0$, and the solution becomes

$$u = \Sigma \sin (n\pi x/l)a_n \cos (n\pi ct/l)$$

When $t = 0$, $u = f(x)$; therefore we must have

$$f(x) = \Sigma a_n \sin (n\pi x/l)$$

But this means representing $f(x)$ by a Fourier series, and we have seen how to solve this problem (i.e. how to determine the coefficients a_n). This completes the solution; the form and motion of the string are now determined at any time.

From music we turn to elementary entities (a step which would have pleased Pythagoras) to examine how far these ideas enter into wave mechanics. We have seen that a wave-group forms an appropriate basis for the mathematical representation of electrons and the like, but it is not immediately obvious how we are to proceed from this point (i.e. how we are to use the system to make mathematical predictions). We know also that the resolution of a complicated situation into simple situations, or the replacement of new facts by arrangements of simpler and more familiar facts, is the surest way of gaining ground in physics. What are the simple and familiar facts about electrons and atoms? We can recognize only the existence of certain *states* (e.g. an atom is identified in spectroscopic analysis by the light it emits when it changes its energy level), and we can think of an atom in a 'ground' state or in an 'excited' state, to each of which corresponds a definite energy level. (This is the basis of Bohr's theory of atoms; the atom emits quanta of radiation as it drops from one energy level to another.) Thus we move to greater and greater abstraction; the atom or any other primary system now is replaced by a *state*. We are obviously very far away from whirls and knots in the æther!

The next step was taken in 1926 when the German mathematician Erwin Schrödinger associated a complex quantity ψ, called the *wave-function*, with the state of the system, and gave the partial differential equation satisfied by ψ. In its simplest form, applicable to a beam of electrons, ψ may be written

$$\psi = Ae^{-int}$$

(which is a wave in complex form) and the amplitude A, which measures the intensity of the disturbance at any point, is to be interpreted thus: A^2 (which is the square of the modulus of ψ), measures the *probability* that an electron will be found at this point. (Thus, in a sense, ψ is a 'wave of probability', but it is a waste of time to attempt to describe abstract ideas by phrases of this kind.) The probability aspect enters for precisely the reasons set out on p. 115, where we used the analogy of a ship at sea.

We can now carry the concept further and find a definite connection with the ideas developed for the plucked string. A system, such as an atom, can exist in certain states, of energy levels $E_0, E_1, E_2, \ldots$ To every one of these corresponds a wave-function $\psi_0, \psi_1, \psi_2, \ldots$ Consider a system in a definite state specified by an energy level $\psi(x)$, where x is a space coordinate. We suppose that we can resolve this state into simpler states in exactly the way that a radio signal is resolved into its frequency composition, that is

$$\psi(x) = a_0\psi_0(x) + a_1\psi_1(x) + a_2\psi_2(x) + \ldots$$

In this, $|a_0|^2$ represents the probability that the energy level E_0 is observed, $|a_1|^2$ the probability that E_1 is observed, and so on. It turns out that the functions $\psi_0, \psi_1, \psi_2 \ldots$ are the eigenfunctions of Schrödinger's equation, and from this follows a fundamental axiom of quantum mechanics, that the only possible values of an observable quantity which can be obtained in an experiment are the eigenvalues of the wave equations of quantum mechanics.

We have travelled so far from simple musical notes to atoms and electrons that it is as well to pause here to look at the matter broadly. What has been done by the mathematicians is to devise a system of calculation which is admirably adapted to describing what is measured in the laboratory when the experiments involve atoms and electrons. It does not matter in the least that we cannot find a name for ψ which will enable the worker to form a definite mental image; ψ is simply the solution of a certain equation. We then find that the mathematics itself takes care of the vagueness of ψ by selecting exactly those values which are of interest. From this follows a fundamental principle in quantum mechanics, relating observables to eigenvalues. An ancient philosopher, whose knowledge of wave-motion was limited to the monochord, might equally well announce triumphantly that the only notes which can be observed (i.e. heard) are those corresponding to the eigenfunctions of the wave equation. The different states of an atom are superimposed exactly as the fundamental and overtones are superimposed in the plucked string. The relative magnitudes of the amplitudes of the overtones, in many ways, indicate the probability that they can be distinguished in the musical note; the amplitudes of the wave-functions indicate the probability that the energy levels will be observed. In both cases the effects of superposition follow from a theorem in pure mathematics: if an equation is linear and has linear boundary conditions, the sum (superposition) of any number of solutions is itself a solution.

The 'Method of Characteristics'

We turn now from the sub-atomic world to examine the part played by waves in problems of high-speed flight. Once again, it is convenient to begin with the vibrating string.

As we have seen, any functions of $x + ct$ and $x - ct$ represent waves travelling in opposite directions, with velocity c, along the axis of x. The general solution of the wave equation is the sum of two such terms. We can regard $x + ct$ and $x - ct$ as composite numbers, something like $x + iy$, but now we are dealing throughout with real numbers. We write, for convenience,

$$\xi = x + ct, \eta = x - ct$$

so that the general solution of the wave equation is

$$u = f_1(\xi) + f_2(\eta) = u_1 + u_2, \text{ say} \qquad \ldots(3)$$

Consider a model (which might be made of cardboard) to illustrate this solution for a string of unlimited length, such as that shown in Fig. 28 (a). The horizontal base of this model

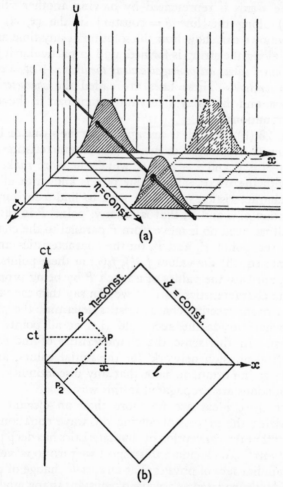

(a)

(b)

FIG. 28. Characteristics for a Vibrating String

represents the plane of x and t, but for convenience the time axis is shown as the length ct. The shape of the wave u_2 at $t = 0$ is a 'silhouette' which can slide on wires parallel to the line $\eta = $ constant in the (x, ct) plane. The travel of the wave $u_2 = f_2(x - ct)$ along the x-axis as time proceeds is represented by moving the silhouette along the wires (which is the same thing as moving the axes to a new origin of time). Similarly, the travel of the wave $u_1 = f_1(x + ct)$ in the opposite direction along the x-axis is represented by moving another silhouette $u = f_1(x)$ along the line $\xi = $ constant in the (x, ct) plane. In this way we can show how the solution at any time and any distance along the string is formed, and in particular it follows that *the initial values are propagated along the lines $x + ct = $ constant, $x - ct = $ constant.* These lines are called the *characteristics of the equation*, and they play a very important part in the analysis of such equations.

In Fig. 28 (b) these ideas are applied to the string of length l (the plane (x, ct) only is drawn). The characteristics are the lines $x + ct = 0$ and $x - ct = 0$, which therefore make angles of 45° with the x-axis. The value of u is known at all points on the characteristics, and if we wish to find u for any other combination of x and ct, that is, at any 'point' P in the (x, ct) plane, all we need do is move from P parallel to the characteristics to the points P_1 and P_2 on the characteristics and substitute into eq. (3) the values $f_1(\xi), f_2(\eta)$ at these points. This shows again how the values of u reach P by being propagated along the characteristics, so that we can say that the values of u along two intersecting characteristics determine the values of u everywhere inside the rectangle (i.e. at all points and at all times). In this sense the characteristics of the equation form a 'transmission network' for the initial values, and it is particularly important to note that any discontinuities of the given functions are propagated in this way.

So far these ideas are no more than an elegant way of summarizing the process of solving the wave equation, but it turns out that the conception of characteristics has deep physical implications. To elucidate this aspect we turn to waves in air. It is a familiar fact of physics that any *small* change of pressure in a fluid is propagated as a wave of constant shape moving at a constant speed typical of the medium. The velocity of propagation is called the *speed of sound* (in air at normal temperature about 1,120 feet per second or about 760 miles per hour). The word 'small' is essential; very violent disturbances, such as those caused by explosions, change shape and move with variable speed.

Consider what happens when a body moves through air (or what is the same thing, air moves past a body) first with a velocity well below that of sound, and secondly, faster than sound. The two cases are illustrated in Fig. 29.

When the body reaches the point A, it will generate a pressure disturbance which travels with the velocity of sound (a) as a spherical wave (like an expanding soap bubble), and this

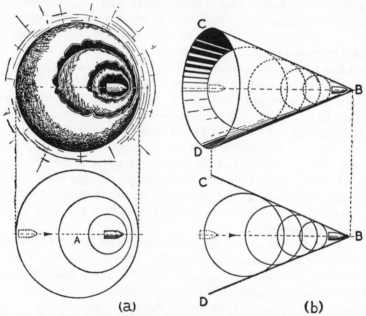

FIG. 29.   Bodies Moving through Air (a) at Subsonic Speed,
(b) at Supersonic Speed

occurs at every point. The pattern of wave-fronts produced by a body moving at a speed $v < a$ is that in diagram (a); no wave meets another, and the body is always inside the pattern. If, however, v is greater than a, the body overtakes the waves, and the pattern is like (b). At any time the waves are contained in a cone of semi angle μ given by

$$\sin \mu = \frac{at}{vt} = \frac{a}{v} = \frac{1}{M}$$

where M is the *Mach number* of the flow pattern, that is, the ratio of the velocity of the body and the velocity of sound. The generators of the cone (in two dimensions the lines *CB* and *DB* of the wedge *CBD*) are called *Mach lines*.

Several interesting features are obvious from this diagram. In the first place, the disturbance caused by the supersonic motion is confined to the cone represented by *CBD*. No disturbance is found outside this cone or, to put it another way, *the air ahead of a body moving at supersonic speed has no warning of the approach of the body*. This is quite unlike what happens with a slowly moving body, with which the pressure disturbances are felt, in varying degree, at all points, so that the air in front of a body moving at subsonic speed has been warned of its approach and has time to get out of the way. Secondly, it is clear that the wave-fronts overlap in a dense fashion on the Mach lines, which thus represent the positions of the most intense disturbance in the fluid. (It is often possible to photograph Mach lines, because of the changes in density which occur along these lines.) The Mach lines are attached to the body, and every point on these lines is moving relative to the air at a speed such that its component perpendicular to the lines equals the speed of sound. The physically important result referred to above is that *Mach lines are the characteristics of certain hydrodynamical equations*.

It is impossible in a book of this type to go into the difficult and involved general theory of characteristics of differential equations, but briefly, we can divide second-order partial differential equations into classes according to their characteristics. Equations with real distinct characteristics are called *hyperbolic*, those with imaginary characteristics, *elliptic*. (The wave-motion equation is a typical hyperbolic equation, and Laplace's equation is elliptic.) The hyperbolic type appears in hydrodynamics only when supersonic velocities occur. A body moving at a speed near that of sound experiences a sudden rise in resistance when the local speed of sound is reached at some point.* The rise reflects the additional energy needed to maintain the wave system attached to the body, and for this reason drag at high speed is separated into *wave resistance* (which can arise in an inviscid fluid) and *frictional resistance*. The outstanding feature of supersonic motion, however, is that

---

* See Chapter 3. The speed of sound depends on temperature and therefore is subject to local variations near a moving body.

it is possible to form patterns of steady flow in which pressure, density, velocity and entropy exhibit *discontinuities*. The discontinuities are confined to the characteristics, or Mach lines. This feature sharply distinguishes the subsonic and supersonic types of motion and the study of high-speed flight is, in many ways, the analysis of highly specialized types of wave motion.

We have seen above that the characteristics may be looked upon as a transmission network for wave patterns. This feature is now widely used in technical problems of high-speed flight to provide a step-by-step method for the numerical treatment of problems which cannot be solved analytically, and in this respect the 'method of characteristics' resembles the

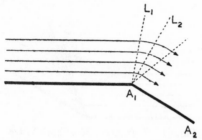

Fig. 30. Supersonic Flow around a Corner (Prandtl–Meyer expansion)

'small arc' process of solving ballistic problems described in Chapter 3.

To understand the basic principles of the process we must first inquire how a high-speed stream turns a corner, such as that shown in Fig. 30.

Suppose that air is moving at supersonic speed over a plane surface with a convex curvature at A_1. When the stream reaches this point there is a sudden small drop in pressure, a pulse of rarefaction, which is propagated as a sound-wave in the region bounded on one side by the Mach line A_1L_1. It can be shown without difficulty that as a result the velocity increases a little across A_1L_1 and the stream is deflected downwards by a small amount. A second drop in pressure produces a second Mach line A_1L_2, inclined at a small angle to the original direction of the stream, and there is another deflection. This process continues smoothly until the corner is turned. Thus a high-speed stream gets over a convex surface by a continuous succession of infinitesimal expansions and deflections, a process known as a *Prandtl–Meyer expansion*. Wavelets of sound spring from the boundary in a diverging pattern, and there is no tendency for one wavelet to pile up on another. On the other hand, if the surface is concave, pulses of compression are produced, so that the successive Mach lines are inclined at greater and greater angles, with the result that the disturbances pile up to form a steep-fronted discontinuity called a *shock wave* (Fig. 31). In passing through a shock wave pressure, density, temperature and velocity jump suddenly to new values.

The step-by-step process of constructing the lines of flow

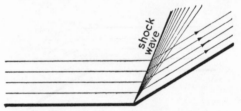

Fig. 31.   Formation of a Shock Wave

rom the characteristics is too involved to be described here in detail, but it is not difficult to understand the basis of the method. In Fig. 32 (a) we suppose that the motion has been completely determined at two neighbouring points P_1 and P_2. Conditions at these points affect only those parts of the fluid lying within the wedges formed by the Mach lines, which intersect at some neighbouring point P_3 if P_1 and P_2 are not too far apart. It is reasonable to expect the flow at P_3 to be uniquely determined by some differential relations holding along P_1P_3 and P_2P_3. P_3 then becomes one of the starting points for the next step. In practice, the characteristics are used to determine a pattern of standing waves of sound, every one of which deflects the flow by a definite small amount (say 2° or 4°, according to the accuracy required). A typical example of the calculated wave pattern and the resulting lines of flow for a high-speed nozzle is shown in Fig. 32 (b). It is interesting to know that these waves have been photographed in real nozzles, and that there is a clear resemblance between the theoretical diagram and the photograph.

We can sum up the use of characteristics by saying that they provide the lines of communication along which the effects of the boundary are propagated into the body of the fluid. Where the network diverges, as in the example of an expand-

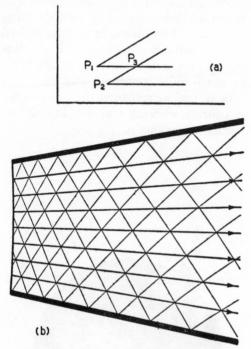

FIG. 32.   The Method of Characteristics

ing nozzle, the process is smooth, but if the wavelets run together, shock waves are formed.   In subsonic flow the effects of the boundary reach the main body of the fluid only by viscosity (i.e. by a process of *diffusion* arising from the incessant agitation of the molecules).   In supersonic flow the boundary affects the main body of fluid chiefly by wave propagation, although viscosity plays a part here also.

☆   ☆   ☆

When Planck devised the quantum theory at the beginning of this century it seemed for a time that the wave had been dethroned and that the particle must take its place. The re-emergence of the wave as a central feature of modern mathematical physics is to be attributed, in part, to the very general nature of the modern definition, but also, perhaps, because the universe is fashioned in this way. With the reappearance of the wave, however, there has come an increasing tendency towards abstraction. The nineteenth-century physicist could not imagine waves without a medium and was thus driven to invent the luminiferous æther. It is probably fair to say that to-day we are much more concerned with what is propagated than what vibrates, and we recognize the wave theory primarily as a calculus, a means whereby man sets in order his thoughts concerning the nature of things.

5

THE MATHEMATICS OF FLIGHT

*In offering to the public the first instalment of the present work, the
author desires to record his conviction that the time is near when the
study of Aerial Flight will take its place as one of the foremost of the
applied sciences, one of which the underlying principles furnish some of
the most beautiful and fascinating problems in the whole domain of
practical dynamics.*

F. W. LANCHESTER, *preface to* Aerodynamics (1907)

The Problem

ALTHOUGH MAN DID not learn to fly until the present century,
the desire to emulate the birds goes back to the beginnings of
history, and there can be little doubt that the myth of Icarus
commemorates an unlucky experimenter of long ago who
found, to his cost, that a mere pictorial resemblance to a bird
is not enough. A primary cause of the failure of early attempts
to fly was the lack of an adequate source of power, a problem
not fully solved until the invention of the internal-combustion
engine but, in addition, the magnitude and nature of the
forces required to sustain a body in the air (the so-called
'secret of flight') were not properly understood until comparatively recently.

An aircraft is a mechanical device capable of sustained and
controlled motion through the atmosphere. It flies either
because it is lighter than air, or because it is so designed that a
large part of the reaction to its motion through the resisting
air is opposed to the force of gravity, or weight, of the whole
machine. Buoyant devices, such as balloons and airships, are
now obsolete except for specialized purposes and need not be
considered further. The modern flying machine is essentially
a specially shaped hollow body equipped with rigid supporting
and stabilizing surfaces (wings and fins), control surfaces
(ailerons, elevators and rudders) and a means of developing

thrust (airscrews or jet-forming devices). Living creatures, which rely on flapping wings, resemble aircraft in many ways, but their flight is more complicated and, as yet, hardly susceptible of mathematical analysis.

A body moving through a fluid (liquid or gas) experiences a resistance. For bodies moving through the atmosphere this

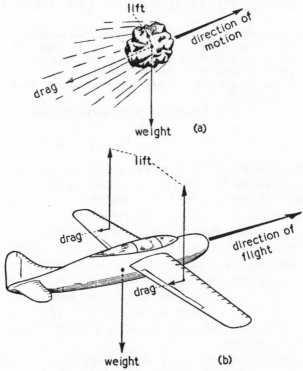

FIG. 33.　Forces on a Body in Flight

resistance is called the *aerodynamic force*. For convenience, we divide this force into two components, called *lift* and *drag*, resolved perpendicular to, and in the direction of flight, respectively (Fig. 33). When a body of irregular shape, such as a pebble, is thrown into the air the atmosphere does little to sustain the flight and acts chiefly to retard the motion. In dynamical terms, the aerodynamic force is nearly all drag with very little lift. The possibility of artificial flight, as known at present, turns on the fact that this state of affairs can be reversed. An *aerodynamic body* is one which, by reason of its shape and finish, is adapted to use in aeronautics, and among such bodies there is a specially important class, called *aerofoils*, which develop much more lift than drag when moved rapidly through the air in a prescribed manner. The most familiar examples of aerofoils are the wings and tail planes of a conventional aircraft.

Modern aerofoils have a wide variety of shapes, to suit the purpose for which they are used, but for aircraft designed to fly at moderate speeds (well below that of sound-waves in air) the generic form of the cross-section of a wing, the so-called *profile* of the aerofoil, is that shown in Fig. 34. The characteristic

Fig. 34. Typical Aerofoil Profile for Subsonic Flight

features are the thick, rounded nose at the leading edge, the smooth, curved upper surface and the tapered tail at the trailing edge. Such a shape resembles the wing of a large bird.

The science of aerodynamics is largely concerned with the calculation of the aerodynamic force set up when such a body is moved through the air, but there is a second class of problems to be solved before an aircraft can be designed, namely, that dealing with stability, or safety in flight. An aircraft, when flying straight and level, may be likened to a see-saw, the principal balancing forces being located in the wings and tail planes. The mathematics, however, is greatly complicated because the motion is in three dimensions, and in addition to pitching (like a see-saw) about a transverse axis an aircraft can rotate about its longitudinal and vertical axes. For safe flight it is essential that the machine be balanced by a proper disposition of forces, but it is also necessary that if a disturbance occurs, the recovery shall be rapid, simple and, as far as possible, automatic (i.e. without conscious intervention on the part of the pilot). These requirements are summed up in the statement that an aircraft must be both *statically* and *dynamically* *stable*. An analysis of the flight of an aircraft is complicated not only because the motion is three-dimensional but also because any disturbance from the trim position involves changes in the aerodynamic forces. We shall not attempt to deal with these matters here.

The physical picture of the action of an aerofoil is not difficult to grasp. In straight level flight the greater part of the weight of the machine is carried by the wings, which are then usually inclined at a small angle to the horizontal. A conventional aircraft differs fundamentally from a bird in that the wings are fixed in the body; there is no flapping motion and the lift arises entirely from the motion of the air past the wings. Clearly, *shape* must be an important factor in the production of lift, and the discovery, in the nineteenth century, that a cambered or arched surface is more efficient than a flat plate in generating lift marked an important advance toward true mechanical flight.

Consider a typical wing, inclined at a small angle to the horizontal, as in an aircraft in straight level flight. In problems of aerodynamics it is often more convenient to think of air sweeping over a body instead of a body moving through still air—in other words, to think of a wing more as a shape suspended in a wind-tunnel than as part of a real aircraft. The intrusion of a body into an air-stream causes local changes in the pressure and velocity of the otherwise uniform air-flow. The pressure of the air affects the wing by producing at every point a force acting perpendicular to the surface, equal in magnitude to the local value of the pressure multiplied by the infinitesimal area surrounding the point, and the aerodynamic force is obtained by summing (integrating) these infinitesimal thrusts over the whole surface. (In addition, there are tangential forces caused by the air sticking to the surface— i.e. by viscosity—which contribute to the drag, but for the present such forces will be ignored.) Experiment shows that pressure near the upper surface of the wing is much less than that near the lower surface, so that the modified pressure field produces a net force on the wing, acting upward (Fig. 35). The component of this force perpendicular to the direction of flight is the lift.

Any body placed in an air-stream causes local perturbations

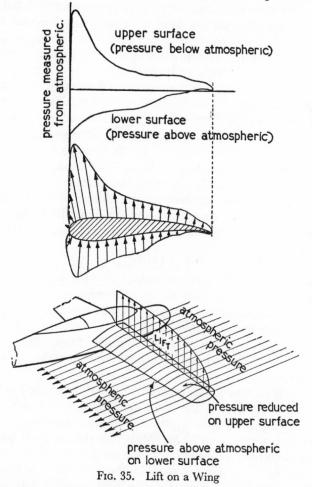

FIG. 35.   Lift on a Wing

of pressure, and a well-designed aerofoil is one which, chiefly by its shape, modifies the pressure field to produce a large lift and a small drag when the aerofoil is inclined at small angles to the direction of the oncoming air. The pressure changes, although greatest in the vicinity of the body, are felt in varying intensity at all points in the air, including the surface on which the atmosphere rests. Thus ultimately the weight of an aircraft is carried by the earth, with the air as the medium which transmits the weight to the ground.

The changes in pressure around the wing are associated with simultaneous local variations in the speed and direction of the air-flow. This indicates another useful way of regarding the production of lift. The air-stream changes its momentum in passing around the body and, by Newton's laws of motion, the body simultaneously experiences a recoil equal to the rate of change of momentum of the air-flow. From this we can gain an insight into the form of the mathematical expression for aerodynamic force. If the density of the air is ρ and its velocity V, in unit time a mass of air ρV passes through unit area perpendicular to the direction of flow. The momentum of this mass is $\rho V \times V = \rho V^2$. A body of cross-sectional area S submerged in this stream will cause a change of momentum of the order of $\rho V^2 S$ in unit time, and this must be proportional to the aerodynamic force. For the lift component it is customary to write

$$\text{lift} = \tfrac{1}{2}\rho V^2 S C_L$$

where C_L is a pure number called the *lift coefficient*. (The reader will recognize that C_L is analogous to C_D, the drag coefficient introduced in Chapter 3.) In general, C_L depends on the shape and attitude of the body, the velocity of the stream and on certain properties of the fluid. The lifting qualities of an aerofoil are then conveniently summarized by the values of C_L.

All this is no more than a crude first analysis of the problem, and the generation of lift by a real wing is a much more complicated process than is indicated above. Nevertheless, the simple arguments given above constitute the essential basis of the mathematical treatment. The outstanding property of an aircraft wing is perhaps not so much the production of high lift, but the fact that for a given wing the lift is uniquely determined by the speed and direction of the relative air-flow, so that, within limits, the amount of lift can be varied easily and smoothly. Without this property it is doubtful if safe flight could be achieved with fixed wings.

The main aim of aerodynamics is to develop such intuitive

pictures into precise physical schemes which can be analysed mathematically, or, in other words, to predict aerofoil properties by calculation. This means replacing what is, in effect, a literary description by exact mathematical statements, a task as yet only partly accomplished. The essential preliminary to such a process is a detailed examination of the fundamental concepts of fluid motion, and this brings us face to face with some of the most difficult problems of mathematical physics.

The Fundamentals
of Fluid Motion Theory

It may come as a surprise to many readers to learn that, at this late stage in the development of mathematics, it is still not possible to calculate exactly the resultant force experienced by a body of any given size and shape moving at a given speed through a fluid. There are, it is true, a few nearly exact solutions for bodies of simple geometrical shape moving at very low speeds, and the mathematical theory has succeeded in predicting many of the outstanding features of forces experienced by bodies of simple shape moving at very high speeds through gases, but apart from these examples, theory must be supplemented by experimental data in any realistic analysis. There is still a large element of empiricism in the treatment of problems of real aircraft.

Like any other branch of applied mathematics, fluid dynamics is an abstraction from physical experience, with considerable idealization in the initial stages. Consider what is seen when a body of simple shape (such as a sphere or a model of an aircraft wing) is suspended in a wind tunnel and the motion of the air is made visible by the introduction of smoke near the tunnel inlet. If the speed of the air-stream is kept fairly low, the picture is one of continuous motion, with the paths of the smoke particles adapting themselves to the contours of the body. A closer examination shows that the individual particles have a wide range of speeds and that there is a tendency to form distinct and enduring patterns of flow. We know, of course, that the continuity, at least, is illusory and that air is really a great mass of molecules with highly individual motions, but, as we have seen in Chapter 2, no significant error can result in problems of aerodynamics if we adopt the hypothesis that air is a continuous medium, capable of being divided without limit into infinitesimal 'fluid particles' (or elementary volumes) which may be supposed to have all the physical properties of fluids in bulk. Such an idealization is necessary if we are to apply the calculus to problems of fluid motion. We can go further than this and consider, instead of a fluid, a *field*, in which physical entities such as velocity, density and pressure have, at any instant, well-defined values at all points.

If, however, we postulate fluid particles we must be prepared to accept that such particles are free to move in the field, and that in doing so, any of their properties, such as density, may change. The change may take place in two ways: (i) because the particle changes its position in the field and (ii) because the field itself is changing with time. Thus statements of the laws of motion of fluid dynamics necessarily take the form of partial differential equations because there are four independent variables, three spatial coordinates and time. In ordinary dynamics time is the sole independent variable and the equations of motion involve only ordinary derivatives.

There are two possible lines of attack on the general problem. The first method, generally known by the name of Lagrange, starts with the concept of a typical fluid particle, subject to the ordinary laws of Newtonian dynamics, moving in the fluid field. The aim of this method is to find a mathematical expression which describes the movements of the particles during a prescribed interval of time.* The second method, initiated by Euler, tries to find an expression giving the magnitude and direction of the velocity at every point in the fluid at any time. The difference between the two approaches is illustrated by the analogy of a horse race: the Lagrangian method amounts to a 'running commentary' on the progress of every horse in the race from start to finish, whereas the Eulerian system provides a sequence of instantaneous photographs of the whole race-course from which the story may be reconstructed. In principle, the two methods are equivalent, but in practice Lagrange's method usually involves greater technical difficulties than Euler's.

---

* Mathematically, to express the position vector of every particle in the fluid as an explicit function of time.

The tendency of fluid motions to form stable patterns emphasizes the important part played by purely kinematical considerations in aerodynamics. Euler's method aims at finding the expression for the velocity vector V as a single-valued function of space and time. At any fixed time it is possible to form an instantaneous picture of the entire motion by constructing curves which have the same direction as V at all points. Such curves are called *streamlines*. If the field is *steady*, that is, if properties measured at a point are independent of time, the streamline pattern is unchanging and may be found experimentally by injecting colouring matter or smoke from a number of points equally spaced across the stream.

Streamline Maps for Ideal Fluids

The concept of streamline patterns is one of the happiest ideas in mathematics and considerably simplifies the process of visualizing fluid motion, but before considering the construction of such patterns, or maps, it is necessary to introduce yet another idealization. Any fluid, whether liquid or gas, is subject to *internal stresses*. When a fluid is at rest the stress between contiguous parts is *pressure*, which always acts perpendicularly to the interface between the parts. When a fluid moves there is relative motion between contiguous parts, and tangential stresses, resembling friction between solids, are set up. The existence of such stresses, expressed by the term *viscosity*, greatly complicates the mathematical treatment, and the study of fluid dynamics, both historically and practically, begins with a hypothetical *ideal* or *inviscid fluid*, which cannot support any tangential stress, however small.

At first sight this particular idealization appears to entail no great break with reality, because the dynamic viscosity of air is very small (about 10^{-4} g/cm sec), and it seems reasonable to assume that such small forces cannot affect the motion seriously. To some extent this is true, but the persistent neglect of viscosity is fatal for the development of fluid dynamics. The reason is that although at some distance from a solid surface the air behaves exactly as if it were devoid of internal friction, in regions adjacent to the body the viscous forces are very important, and the ultimate pattern of flow of the fluid depends largely on what happens in the so-called *boundary layer*, a very thin sheath of slowly moving air which envelopes the entire body. Actually, the peculiar property of aerofoils of producing a powerful lift can be traced to viscosity, and mechanical flight would be impossible if the atmosphere were truly devoid of friction. We shall return to this important point later.

However, the study of ideal fluids is especially appropriate as an introduction to *two-dimensional aerofoil theory*, which gives a realistic account of lift, but affords no basis for the estimation of drag, a defect which is to be ascribed to the assumption of zero viscosity. In considering motion past a body which is very long across stream, it is reasonable to suppose that near midstream the motion is independent of the *exact* position of the

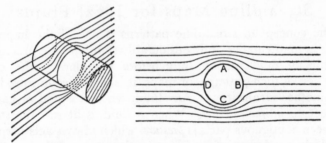

FIG. 36.   Flow Past a Long Circular Cylinder

reference plane parallel to the direction of flow and, further, has no component perpendicular to this plane. (Physically, this means ignoring all effects at the tips of the body.) Such motion is called two-dimensional, and in this case the streamlines are plane curves. In three-dimensions, the concept of streamlines leads easily to that of *streamtubes*, formed by taking a closed curve in the fluid and drawing the appropriate streamlines through every point on the curve to form a surface. The space occupied by the moving fluid may be imagined to be filled with such tubes, with fluid moving parallel to the walls, as in real tubes. The two-dimensional streamline pattern then becomes a plan view of such an assembly of tubes. Fig. 36 shows the streamline map for the simplest type of two-dimensional motion of an inviscid fluid past a long circular cylinder.

If the circumstances of the motion are such that variations

in the density of a moving fluid particle are unimportant, the fluid is said to be *incompressible*. Since matter cannot be created or destroyed by motion, the mass of fluid which enters any length of streamtube must be the same as that which leaves, and therefore, by the arguments of p. 52, the speed of an incompressible fluid is inversely proportional to the spacing of the streamlines. A *properly constructed* streamline map thus indicates how the direction of flow changes, and also reveals changes in speed. In Fig. 36 the streamlines are equally spaced at great distances from the cylinder, indicating that in these regions the fluid is moving everywhere with the same speed, but at the points A and C, where the streamlines are closely packed, the fluid is accelerated to speeds considerably above that of the undisturbed stream.

The principle of conservation of mass (or of *continuity*, as it is called in fluid dynamics) provides the key to the mathematical representation of two-dimensional streamline maps. Suppose that a particular streamline has the equation $\psi(x, y) = C$, where C is a constant and x and y are ordinary cartesian coordinates. Because a streamline, by definition, is everywhere parallel to the direction of motion of the fluid there can be no flow across it. If we accept this as the fundamental physical property of a streamline we can try to express the idea mathematically by taking the difference between the values of $\psi(x, y)$ at *any* two points to be proportional to the mass flux across a curve joining these points. (This at least is consistent with the definition of a streamline, for along such a line $\psi(x, y)$ has the same value C and therefore zero difference.) This supposition turns out to be correct. By combining the principle of conservation of mass with the geometrical definition of a streamline it can be proved that there exists a function $\psi(x, y)$, called the *streamfunction*, which has the following properties:

(i) the difference between the values of ψ at any two points in the fluid gives the rate of transport of fluid across *any* curve joining the points;

(ii) the velocity components are given by

$$u = -\frac{\partial \psi}{\partial y}, \quad v = \frac{\partial \psi}{\partial x}$$

and

    (iii) the streamlines can be plotted by giving the constant
         C different values in the equation $\psi(x, y) = C$.

The 'proper method' of drawing a streamline map referred to above is, in fact, simply that of giving C equally spaced values (say $C = 0, \pm 1, \pm 2, \pm 3, \ldots$, according to the scale chosen), so that equal amounts of fluid flow in equal times in the space between adjacent streamlines. The distance apart of any two streamlines is then inversely proportional to the mean velocity of the incompressible fluid in the region between the streamlines.

A streamline pattern or map simply shows how incompressible fluid can move without violating the principle of conservation of mass. If the streamfunction ψ can be found, the velocity at any point is also determined, so that to make further progress we must now investigate methods of finding ψ.

Patterns of Irrotational Motion

Fluid motion is characterized by an embarrassingly wide range of possible modes, and some restriction must be imposed if the analysis is to remain simple. If we consider the general motion of a fluid particle typified in two-dimensions by, say, a square, the following changes are possible during a brief interval of time:

    (i) the element is *translated*, that is, its centre has moved a
       certain distance;
    (ii) the element is *rotated*, that is, an axis of symmetry,
       such as a diagonal, has turned about the centre;
    (iii) the element has undergone *strain*, that is, the lengths
       of the sides have changed, the area remaining the
       same.

These changes are illustrated in Fig. 37 (a).

The simplest type of motion is one in which the elements undergo translation and perhaps distortion (strain) but do not rotate A motion of this type is called *irrotational*. This definition does not affect in any way the ability of the fluid particles to follow curved paths, and it is thus quite logical to speak of a finite mass of fluid rotating irrotationally around a body.   The simplest example of this kind of motion is afforded by a giant wheel at a fun-fair—the cars, which always remain upright (at least approximately), correspond to fluid particles moving irrotationally about a centre.

With the restriction to steady irrotational motion of an incompressible inviscid fluid, the two-dimensional kinematics of fluids takes a very simple form.   In the first place, it can be proved that in these circumstances there always exists a function

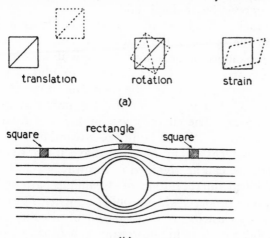

FIG. 37.   (a) Motion of a Fluid Element (b) Irrotational Motion

$\phi(x, y)$, called the *velocity potential*, such that if u and v are the components of velocity in the x and v directions, respectively,

$$u = \frac{\partial \phi}{\partial x}, \qquad v = \frac{\partial \phi}{\partial y}$$

Conversely, if a velocity potential exists, the motion must be irrotational.

So far we have dealt entirely with real variables.   The complex variable $z = x + iy$ (Chapter 2) may also be used to specify position in the (x, y) plane.   Among the possible functions of z there is one important class distinguished by the name holomorphic *;   these are single-valued functions which are 'well behaved' in the sense that they have no discontinuities, such as infinities.   Thus $z, z^2, z^3, \ldots$, $\sin z, e^z$, or any finite combination of them, are holomorphic;   the functions $1/z$,

* The words *analytic* and *monogenic* are also used.

log z, ... are holomorphic except at $z = 0$, where they become infinite. This selection of functions leads to one of the most striking facts in mathematics, that *the velocity potential ϕ and the stream function ψ are always the real and imaginary parts of a holomorphic function of the complex variable $z = x + iy$.* Conversely, the real and imaginary parts of any holomorphic function of z are the velocity potential and streamfunction, respectively, of a possible two-dimensional irrotational motion of an incompressible inviscid fluid.

The reader will see at once that this places a powerful weapon in the hands of the mathematician. He has only to choose a holomorphic function and split it into its real and imaginary parts (a very simple task) to obtain a possible streamline pattern. Thus the function $z^2 = (x + iy)^2 = x^2 - y^2 + 2ixy$ has $x^2 - y^2$ for its real part and $2xy$ for its imaginary part. The curves given by $2xy = $ constant, which are hyperbolas, are then a family of streamlines. An infinite variety of patterns, many of which have interesting physical interpretations, can be obtained in this way.

The physical interpretation of streamline patterns depends primarily on a property which would be possessed by an inviscid fluid, if such existed. A fluid devoid of viscosity could slip over a solid surface without restraint, because there are no tangential forces at the boundary. (Actually, no real fluid can do this, because viscosity prohibits relative motion at the interface of a solid and a fluid.) Since fluid cannot cross a streamline, any curve of the pattern may represent a solid (impervious) surface in the fluid without violating any property of streamlines. Further, any region of space inside a closed streamline, or wholly on one side of a streamline, may be supposed occupied by solid matter and the remainder of the space by fluid, without disturbing the pattern in any way. Thus in Fig. 36 the circle which is supposed to represent the cross-section of the cylinder is actually part of the streamline $\psi = 0$ (the remainder of the streamline is the x-axis), and any pattern inside the circle may be ignored.

The streamline map for irrotational motion past a body of simple shape may thus be constructed by finding, by intuitive methods, a holomorphic function of the complex variable z called the *complex potential*) such that its imaginary part gives rise to a family of curves, one of which must conform to the contour of the body. For simple figures this process is not too difficult, but if the mathematician is not fortunate enough to 'spot' the correct complex potential, there is another way in which the problem can be solved. It is easily shown that both ϕ and ψ satisfy Laplace's equation $\nabla^2\phi$ (or $\nabla^2\psi$) $= 0$ (p. 65), and the determination of the pattern amounts to finding the solution (of either equation) which satisfies certain boundary conditions. This type of analysis is often difficult and, fortunately, unnecessary for the limited purposes of aerodynamics. It is simpler and more convenient to start with a very easy problem, that of irrotational motion past a circular cylinder, and from this to build up the streamline pattern around less simple shapes, such as wing profiles, by a geometrical method which does not involve the difficult process of solving a second-order partial differential equation. This is the method now universally followed in teaching aerodynamics.

The Dynamical Problem and its Solution

All that has been discussed so far is geometry, and we have now to deal with the dynamical problem of calculating aerodynamic force, which, as we have seen, means finding the disturbed pressure field near the body. This can be done directly from solutions of the equations of motion for appropriate boundary conditions, but the assumption of steady two-dimensional irrotational motion, combined with the kinematical approach, allows the problem to be solved much more easily.

Pressure (which in a moving fluid may be treated as the sum of two parts, *static pressure*, which persists from the state of rest and *dynamic pressure*, which arises from the motion) may be regarded as a form of energy. Kinetic energy, generated by the motion of the fluid, is expressed, per unit mass of fluid, by $\frac{1}{2}\rho V^2$, where V is the resultant velocity. The principle of conservation of energy, when applied to fluids, is known as *Bernoulli's theorem*, which states that *in steady irrotational motion (not subject to external forces), the sum of the pressure and kinetic energies, per unit mass of fluid, has the same value at all points in the field.* The presence of a body in a stream causes energy to be redistributed between the two forms, so that where pressure is high, velocity is low, and vice versa.

Bernoulli's theorem supplies the essential link between the kinematical and dynamical studies. Consider a body held at rest in a steady incompressible stream of air. Far away from the body (mathematically, at infinity), the pressure (p) and the velocity (V) have the same known fixed values (p_0, V_0) at all points. Bernoulli's theorem states that

$$\underset{\text{(near the body)}}{\tfrac{1}{2}\rho V^2 + p} \; = \; \underset{\text{(far from body)}}{\tfrac{1}{2}\rho V_0^2 + p_0} \; = \; \underset{\text{(all parts of the field)}}{\text{constant}}$$

Thus if V is known from the kinematical study, p is easily found at all points, and in particular along the streamline which represents the profile of the aerodynamic body. The net thrust of the fluid on the body (the aerodynamic force) can then be found by integration of p over the profile. Thus, in principle at least, one of the main problems in aerodynamics is solved. Schematically, the successive steps in the solution are as follows:

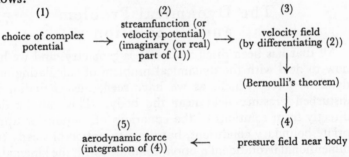

In this way a two-dimensional aerodynamic problem becomes, in effect, a problem in kinematics (geometry), with the formidable equations of motion appearing only at the final stage, disguised as Bernoulli's theorem. Let us see what this gives in the simple problem of a long cylinder placed across a steady stream (Fig. 36). It is not difficult to find, by trial and error, a complex potential with an imaginary part which yields one streamline consisting of a straight line and circle, and a uniform flow (straight parallel streamlines) at infinity. The subsequent steps, calculation of the velocity and the pressure (by Bernoulli's theorem) and integration of the pressure thrusts over the circle, offer no particular difficulty.* The final result, however, comes as something of a shock—the total thrust on the cylinder (the aerodynamic force) is exactly zero! If this were true, a long, round stick placed across a stream would need no restraint to hold it in position.

There is no mistake in the mathematics, and the reader can see for himself, by looking at Fig. 36, that the result is inevitable. In general terms, Bernoulli's theorem implies that the pressure field is a kind of 'negative' (in the photographic sense) of the velocity field (pressure is low where velocity is high, and vice versa), so that if the streamline pattern is symmetrical, the pressure field also must be symmetrical and there can be no *net* force on the cylinder. The result illustrates a more general theorem, known as d'Alembert's Paradox, that there can be no *drag* in an inviscid fluid. To hold a body of any shape at rest in a stream requires a force and a couple, but the component of the force in the direction of the stream (i.e. the drag) is always exactly zero. The result is caused by the removal of viscosity from the calculations.

D'Alembert's Paradox was long regarded as indicating that the dynamics of a frictionless fluid could play no part in problems of flight but, fortunately, this is not so. The theorem does not exclude the possibility of calculating lift and, happily, the inviscid fluid theory is able to do this with a very fair measure of success. In the problem of the cylinder, both lift and drag are zero because there is complete symmetry. Is there any way of introducing asymmetry into the streamline pattern so that lift, at least, does not vanish? It turns out that we can do this without difficulty, and the investigation provides, at the same time, the long-sought essential clue to the production of lift—the essence, in fact, of the 'secret of flight'.

Circulation and the Origin of Lift

Consider (Fig. 38 (a)) fluid rotating irrotationally around a circular cylinder, but so that the velocity at any point is inversely proportional to distance from the centre. (This is a special (cyclic) type of motion, not to be confused with a mass

* See Appendix VII. The complex potential is simply $V\left(z + \dfrac{a^2}{z}\right)$, where a is the radius of the circle.

of fluid rotating like a solid body.)   The streamlines of this motion are concentric circles.   If this pattern is combined with that of Fig. 36, the streamlines (in certain circumstances) have the shape shown in Fig. 38 (b).   The combined pattern is symmetrical about the line *PQ*, so that there is no net force parallel to the undisturbed stream (i.e. no drag, in accordance with d'Alembert's Paradox), but there is asymmetry about the diameter *RS*, since velocity is high near *P*, where the component motions reinforce each other, and low near *Q*, where the two streams are in opposition.   Thus there is a pressure difference between *Q* and *P* and a force exists in the direction *QP*, perpendicular to the direction of the stream far from the

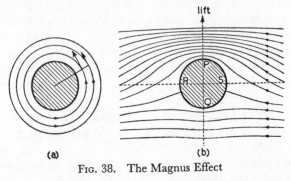

(a)   (b)
FIG. 38.   The Magnus Effect

body.   This example of lift, called the *Magnus effect*, explains the swerve of a 'cut' tennis ball or of a 'sliced' ball in golf.

Neither the cyclic motion nor the simple streaming motion can produce lift on their own, but only in combination.   Is there any simple quantity which can be said to characterize the cyclic motion?   Velocity, being inversely proportional to distance from the centre, depends on the particular circle chosen as streamline, but if velocity is integrated around a circle, the result is a quantity independent of position.*   This constant quantity is called the *circulation*.   Calculation of the aerodynamic force on the cylinder by the method of the previous section shows that the lift per unit length of cylinder is equal to the product of the density of the fluid,

* Thus if $v = K/r$, where r is the radius of a streamline, $\int vds$ (ds = element of arc) is simply $K\int_0^{2\pi} \frac{1}{r} \cdot rd\theta = 2\pi K$, which is independent of r.

the undisturbed uniform velocity (velocity at infinity) and the circulation.

This is a special case of one of the most famous theorems of aerodynamics—the *Kutta–Joukowski theorem*, which may be called the central result of aerofoil theory. *A cylinder of any cross-section at rest in a uniform inviscid two-dimensional stream of velocity V, with a circulation of intensity K around it, experiences a lifting force of magnitude ρVK per unit cross-stream length perpendicular to the direction of V.* The circulation is defined as the line integral of the tangential component of the velocity along any closed curve around the body, and may be shown to be independent of the particular curve. Lift is thus essentially related to circulation around a body in a wind, and the aerodynamical problem is now reduced to that of investigating patterns of flow which give (i) a uniform stream at infinity and (ii) a motion with circulation around an aerofoil profile in the finite region of the plane of motion. The 'secret of flight' is essentially that of finding shapes which give strong circulation and small drag when moved rapidly through the air.

The result given above is a remarkable example of the power of mathematics, for circulation is essentially a mathematical concept which is difficult to explain in other terms. This is not so for other physical entities—momentum, for example, is readily envisaged as the 'weight of a blow'—but circulation does not lend itself easily to this kind of analogy. Approximately, circulation is equal to the mean value of the velocity on a closed curve multiplied by the length of the curve, but this does not suggest why it is fundamental in the analysis of lift. The significance of circulation in the problem of lift arises from its relation to vorticity, or spin, which is explained later (p. 160).

The analysis does not explain how circulation is generated or varied, and this question cannot be answered without bringing in viscosity. For the present, circulation must be accepted as a condition which can exist * in an inviscid fluid but which cannot be started from rest by any combination of forces. This conclusion is drawn from a celebrated theorem proved by Kelvin, that circulation around a closed material

---

* That is, the assumption of such a condition does not involve a logical contradiction.

curve (i.e. a curve which always consists of the same fluid particles and therefore moves with the fluid) cannot vary in an inviscid fluid if the external forces are derived from a potential and density is a function of pressure only. Hence circulation cannot be produced from a state of rest by the motion of an inviscid fluid (subject to the above conditions) in a region surrounded by closed material lines, and this would seem to preclude the possibility of starting a flight. Viscosity comes to our aid here, for although initially the motion of the air around the wings of an aircraft beginning to move off is free from circulation and there is no lift, the action of viscosity quickly forms a small region of rotary motion, called the *starting vortex*, behind the tail. Kelvin's theorem then requires that a circulation of equal strength, but opposite in sign, be developed around the wing as a whole, and it is this circulation which is associated with the lift.

The highly abstract nature of circulation is well illustrated by the fact that F. W. Lanchester, the British engineer who was the first man really to understand mechanical flight, did not use the concept explicitly in his pioneer work. Instead he wrote, somewhat vaguely, of a 'supporting wave'. It was left to the German Kutta and the Russian Joukowski to state the fundamental theorem in its true mathematical form. Had Lanchester been able to put his ideas clearly in mathematical language, there is little doubt that his genius would not have had to wait so long for recognition.

The Kutta–Joukowski theorem does not determine the lift absolutely, since the amount of circulation is left arbitrary and, in particular, unrelated to the speed of the wind (velocity at infinity). In the illustration of the Magnus effect the circulation appears as the constant which determines the speed of the cyclic motion and thus, in theory, may have any value. Experiment shows that the lift of a given wing is uniquely determined by two factors—speed through the air and inclination to the direction of flight. To render the problem determinate some relation must be established between the circulation around the wing and the speed of the relative wind. This brings us to the final step in the development of two-dimensional aerofoil theory, the pioneer work of the Russian scientist Joukowski between 1906 and 1910.

Joukowski's Solution of the
Two-dimensional Aerofoil Problem

The development of the two-dimensional aerofoil theory now falls into two parts: (i) the purely geometrical problem of finding the streamline pattern of motion of an incompressible inviscid fluid around shapes of interest to the aircraft designer, and (ii) the dynamical problem of expressing the lift in terms of quantities easily measured or estimated by the engineer. These problems were effectively solved by Joukowski in the early years of this century.

The first problem was treated by a method analogous to that

FIG. 39.   A Wing as a Distorted Circular Cylinder

employed in making a wing-shape in modelling clay. This clay is usually supplied in small circular cylinders. It is clear from Fig. 39 that the nose of a typical wing is not unlike part of such a cylinder, and that the tail could be constructed by pulling out the clay into a tapering section—in other words, a typical aerofoil profile can be modelled from a circle by a process of *continuous distortion*. Can this be represented mathematically?

The process of distorting a circle into an aerofoil shape can be done by a sequence of purely geometrical operations, but the complex variable supplies the easiest approach. The simplest type of fluid motion is that represented by a system of parallel, equally spaced straight streamlines, every one of which has the equation $\psi =$ constant (Fig. 40). The equipotential lines $\phi =$ constant cut these at right angles, and the resulting mesh can be regarded as a coordinate system, in which any point is determined by a pair of numbers (ϕ, ψ). Equally, the $\phi\psi$ plane may be regarded as an Argand diagram (p. 31) over which the complex variable

$$w = \phi + i\psi$$

takes different values. We call this diagram the w-plane.

Suppose now that we put $w = f(z)$, where $z = x + iy$ is a new variable which determines the position of the point (x, y) on what we shall call the z-plane and f is an arbitrary function. Let $f(z)$ be a holomorphic function of z. Then

$$f(z) = \phi + i\psi$$

so that ϕ and ψ become the real and imaginary parts of a holomorphic function of z, and therefore must be the velocity potential and the stream function, respectively, of a possible

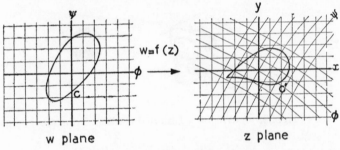

w plane   z plane

Fɪɢ. 40.   Conformal Mapping

two-dimensional irrotational fluid motion in the z-plane, but, in general, such a motion will be more complicated than that in the w-plane. A curve C in the w-plane (i.e. an assembly of points (ϕ, ψ)) is likewise changed into a curve C' of different shape in the z-plane. A process of this type (called a *transformation*) amounts to making a map of the w-plane in another system of coordinates (just as the spherical surface of the earth is mapped on a plane by Mercator's projection), and the function $f(z)$, which connects the two planes, is called the *mapping function*. At the same time $f(z)$ can be regarded as the complex potential of a fluid motion in the z-plane and, combining the two ideas, we may look upon the complex potential $f(z)$ as the mapping function which changes the simple uniform rectilinear flow parallel to the ϕ-axis in the w-plane into a more complicated flow pattern in the z-plane.

Transformations effected by holomorphic functions are called *conformal*. This means: (i) the angle between two curves passing through a point in the *w*-plane is equal to the angle between the corresponding curves in the *z*-plane and (ii) corresponding infinitesimal elements of arc are magnified or diminished in the ratio $|dw/dz| : 1$. Thus dw/dz measures the amount by which an infinitesimal arc δw is rotated and lengthened or diminished in changing into an infinitesimal arc δz on the *z*-plane.

In fluid dynamics, conformal transformation provides an easy and systematic way of proceeding from simple to more complicated flow patterns. In aerodynamics, the process is particularly valuable because there is a general class of mapping functions (known as *Joukowski transformations*) which distort circles into shapes of greater interest to the engineer, leaving the flow pattern at infinity unaltered. In this way a more

FIG. 41.   Shapes Obtained from a Circle by Joukowski Transformations

complicated streamline pattern can be formed from a simple basic pattern. In aerodynamics the starting point is irrotational motion with circulation past a circle (Fig. 38), and by the use of a very simple mapping function Joukowski produced a variety of distorted or degenerate circles, such as ellipses, pear-shaped curves, straight lines, circular arcs and finally a closed curve with a thick rounded nose, smooth arched upper surface and a sharp tail (Fig. 41). The last-named shape resembles the profiles used by the early designers for the wings of their aircraft, and is generally known as a *Joukowski profile*. The streamline pattern also changes to conform with the new shape, except in regions remote from the profile ('at infinity') so that the details of the motion around the derived shapes are also known.*

Joukowski's method means that in one sense the basic problem —that of finding the streamline pattern of irrotational flow with circulation around a shape proposed by the designer—has been *inverted*. The process produces the streamline pattern for flow past a shape which the engineer can accept as having most, if

* See Appendix VIII.

not all, of the features of a real wing section. It is extremely difficult to determine the circulation around a given arbitrary shape from the equations of motion, and Joukowski's method is a very effective way of avoiding trouble. Since Joukowski's day the method has been extensively studied and extended by Kármán, Trefftz, Mises, Carafoli and others, with the result that profiles of aerofoils having almost any desired features can now be obtained from a circle.

The second problem is solved by a remarkable hypothesis. The outstanding features of a real wing are: (i) ability to set up a strong circulation without mechanical movement of its parts, and (ii) the fact that the lift is uniquely determined by the magnitude and direction of the motion of the air relative to the aerofoil. With a rounded body (such as a circular cylinder) in a uniform wind the circulation can be given any value, and to every value there corresponds a possible irrotational motion, so that the pattern of flow is not uniquely determined by the velocity of the undisturbed stream and the dynamical problem is indeterminate. On the other hand, experiment shows that a cylinder with a sharp trailing edge produces a well-defined lift which depends chiefly on the magnitude and direction of the airflow. Joukowski's transformation distorts the circle into a shape with a single sharp point (a cusp), corresponding to the tapered trailing edge of a real wing. In general, the presence of a cusp implies an infinite discontinuity in the velocity field, but Joukowski proved that for a profile which possesses one such sharp point there exists one motion for which there is no discontinuity at the cusp. *Joukowski's hypothesis* is that this pattern alone is significant, and that the circulation must have exactly that value which is consistent with a finite velocity at the cusp. This condition makes the dynamical problem determinate.

To see what this means, consider what happens with a rudimentary aerofoil represented by a single line (i.e. a semi-infinite lamina). Fig. 42 (a) is actually a transformation of Fig. 36, flow without circulation. The effect of the pressure of the stream is to rotate the lamina clockwise, but there is complete symmetry and hence no resultant force. In Fig. 42 (b) circulation has been introduced with the value given by Joukowski's hypothesis. The fluid leaves the lamina smoothly at the trailing edge, there is well-defined asymmetry and a resultant force at right angles to the undisturbed stream. This is the lift, and in a properly designed aerofoil the circulation automatically takes the value which causes the air to flow smoothly off the tail.

In Joukowski's theory, the unique value of the circulation K which imposes a finite value on the velocity at the cusp is proportional to the undisturbed velocity V. Thus the lift, which by the Kutta–Joukowski theorem is ρKV per unit cross-stream length, is proportional to V^2. Hence C_L, the lift coefficient, is independent of V and depends only on the

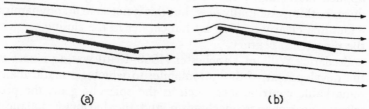

(a)   (b)

FIG. 42.   Rudimentary Aerofoil (a) without and (b) with Circulation

inclination of the aerofoil to the undisturbed stream. The exact relation obtained by Joukowski is

$$C_L = 2\pi \sin \alpha \simeq 2\pi\alpha$$

where α, assumed small (<0.2 radian $\simeq 12°$) measures the incidence of the aerofoil. This theoretical result is in good agreement with wind-tunnel measurements of the lift of simple wings of large span.

The two-dimensional aerofoil theory was initiated by Kutta in 1902, but it was not until 1906 that Joukowski (who had discovered independently the famous formula ρKV for lift) proved that circulation can be uniquely determined for a profile with a blunt nose and a sharp tail. In 1910 he gave the correct expression for C_L for a 'Joukowski aerofoil'. This achievement marks the emergence of aerodynamics from the highly unrealistic classical hydrodynamics. Undoubtedly, the connection between circulation and lift had been grasped long before this by F. W. Lanchester, and is implicitly contained in his unorthodox *Aerodynamics* published in 1907, but his conclusions were not expressed in conventional mathematical terms and were imperfectly understood by the majority of those who read the work when it was first published.

Lanchester's Theory of the Finite Aerofoil

The analysis of lift given above cannot yet be applied to a real aircraft, since it refers only to wings which extend indefinitely across the air stream.* In other words, the treatment ignores all effects at the wing-tips. In many branches of applied mathematics the inclusion of such 'edge effects' introduces nothing essentially new, but this is not so in aerodynamics, a fact first clearly realised by Lanchester in the closing years of the nineteenth century.

Frederick William Lanchester (1868–1946) was a pioneer in the development of the automobile, to which he made many remarkable contributions, such as the epicyclic gear, the pre-selector gear-change mechanism and the harmonic balancer (to quote only a few). In 1894, nine years before the Wright Brothers flew, he gave the correct explanation of the action of a real wing in a lecture to a local society. His subsequent efforts to get his theory published in a scientific journal failed and he fell back on the unusual plan of disclosing his ideas in book form. Between 1907 and 1908 he produced two large volumes entitled *Aerodynamics* and *Aerodonetics*, respectively. These works may be said to have played a part in aerodynamics not unlike that exercised by Newton's *Principia* in astronomy.

Lanchester's books are not mathematical treatises in the accepted sense of the word, but the writings of an exceptionally gifted engineer who either could not, or would not, express his conclusions in the usual mathematical form. They are examples of 'intuitive' reasoning in the best sense of the word and therefore, at times, difficult to comprehend. To understand Lanchester's attitude to mathematics it must be realized that by the end of the nineteenth century hydrodynamics had become very much a preserve of mathematicians. It was

---

* That is, to wings of *infinite aspect ratio*. The aspect ratio is defined as the number span ÷ chord, where the span is the distance between the wing tips and the chord measures the length of the wing in the direction of motion.

largely ignored by practical engineers, for whom results such as d'Alembert's Paradox and concepts such as inviscid fluids were mere mathematical phantasies. Navier and Stokes had introduced viscosity into hydrodynamics many years before, but the non-linearity of the equations of motion meant that most of the problems of direct interest to engineers were insoluble. This condition of stalemate was not resolved until 1906, when the German mathematician Prandtl introduced the concept of the boundary layer (see p. 140). From that time onwards it has been possible to calculate, with adequate precision, the skin-friction, or tangential force set up by molecular attraction at the junction of a fluid and a smooth solid. This achievement went a long way to make fluid dynamics realistic, but the exact calculation of the complete drag is still beyond the power of mathematics.

A science often seems to progress in a series of well-marked steps, each of which turns the flank of some insurmountable obstacle. Prandtl's introduction of the boundary layer is a good example of such a 'discontinuity'. Although Lanchester clearly knew nothing of Prandtl's work, he had reached essentially the same conclusion and, by the simplest algebra, had deduced formulæ which are very good approximations to the exact results, obtained some years later by another German mathematician. Here, as elsewhere, Lanchester could not reap the full reward of his acute physical insight because he was not familiar with the concepts of mathematical physics.

Lanchester's most profound and original contributions, however, were in relation to the problem of lift. He does not seem to have known of the Kutta–Joukowski theorem and he made no direct use of the mathematical concept of circulation. To explain how a wing lifts, he considers first a kind of parachute sinking under its own weight. In the process, any air dragged down is compensated by an upward flow around the edges, the motion being maintained by the difference in pressure between the upper and lower surfaces. When the parachute is replaced by a glider moving horizontally, the relative motion of the air must be successively upwards, downwards and finally upwards again. Any motion imparted to the air by the wing is given back, so that an aerofoil of infinite span may be looked upon as a device which imposes on the air a kind of wave motion, in which energy is conserved. For this reason Lanchester called the flow about an aerofoil a 'supporting wave'—mathe-

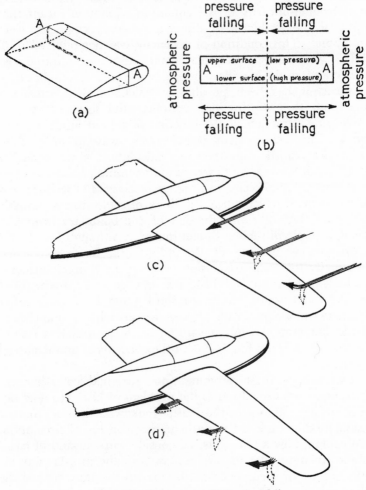

FIG. 43.   Flow over a Finite Aerofoil

matically, a circulation is set up. With a finite aerofoil the argument must be modified—energy is not conserved, because of the effects at the tips, and Lanchester, in this way, reached the important conclusion that a real wing would require a force from without to sustain it—in other words, there must be a continuous supply of energy from an external source (the engine). This kind of motion he called a 'forced wave'. The result is one of the most important in aerodynamics and was the first recognition of what we now call *induced drag* or resistance arising solely from lift, and unrelated to viscosity.

Lanchester next described a possible mode of motion around a finite wing which embodied these ideas. Consider, as in Fig. 43 (a), a section of an aerofoil along the span. Pressure is below atmospheric on the upper surface and above atmospheric on the lower, but equal to atmospheric just outside the wing-tips. Hence pressure must decrease in the direction of the arrows, and since a fluid always tends to move down a pressure

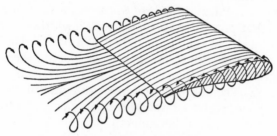

FIG. 44.   Formation of the Tip Vortices

gradient, it follows that an air particle originally moving perpendicular to the leading edge will be deflected as in diagram (c), inwards on the upper surface and outwards on the lower surface. The two airstreams must unite at the trailing edge, so that particles which reach the same point on this edge will have the same speed but will be moving in different directions (d). This means that there is a discontinuity of velocity on a sheet of air extending behind the trailing edge, but this cannot persist for any great distance and the sheet soon rolls up into two columns of rotating air, called the *tip vortices* (Fig. 44). The existence of these vortices has now been confirmed by photographs, but such evidence did not exist when Lanchester evolved his theory.

Lanchester was able to make some striking and essentially correct deductions from this intuitive picture, but the incorporation of the concept of the trailing vortex pattern into a mathematical system was primarily the work of Prandtl at a much later date.

To explain Prandtl's method necessitates the introduction here of one of the most important idealizations of hydrodynamics. If we consider again the cyclic motion depicted in Fig. 38 (a), p. 148, it is evident that the cylinder which is represented by one of the concentric circles is unnecessary for the description of the motion since any circle may be chosen for the purpose. However, if we remove the cylinder, there must be a discontinuity of velocity at $r = 0$, since the velocity at any point varies as $1/r$, where r is distance from the common centre of the circles. We describe this state of affairs by saying that there is a *line vortex*, perpendicular to the plane of the paper, passing through the point $r = 0$. The motion of the fluid around $r = 0$ is called the *induced field of the vortex*. Such motion resembles that popularly associated with a whirlwind, or is like the flow of water at the drainage hole of a bath. The induced velocity decreases with distance from the vortex, the influence of which is effectively restricted to a relatively small region of the fluid. The counterpart of this is also found in nature (e.g. the damage caused by a tornado is usually confined to a small region near the path of the 'funnel').

To enable us to carry out calculations with vortices, we need to consider the analytical expression for the rotation of a fluid particle. In two-dimensions, the angular velocity of an infinitesimal element about an axis perpendicular to the xy plane is

$$\frac{1}{2}\left(\frac{\partial v}{\partial x} - \frac{\partial u}{\partial y}\right)$$

where u and v are component velocities. We call $\frac{\partial v}{\partial x} - \frac{\partial u}{\partial y}$ the *vorticity* of the motion; an irrotational motion is one with zero vorticity. The product of the vorticity and the infinitesimal area $dxdy$ can be shown to be the differential of the circulation (dK), so that alternatively we may define vorticity as the limit of the ratio

$$\frac{\text{circulation around an infinitesimal area}}{\text{infinitesimal area}}$$

as the area is indefinitely diminished.

This establishes the important connection between vorticity

and circulation. A *vortex element* is an infinitesimal area for which the vorticity does not vanish, and the *strength* of a vortex element is defined to be the circulation around it. We may now ignore entirely the cylinder in the motion of Fig. 38, and say simply that the flow is that associated with a line vortex of strength K at the origin, where K is the circulation. Finally, it can be shown that the circulation around a closed curve is equal to the sum of the strengths of the vortex elements contained within it.

All motion of a real fluid is rotational, to greater or less degree, and irrotational motion, or motion entirely free from vorticity, is primarily a mathematical concept. It is impossible to imagine vorticity arising from any circumstance of the motion of an ideal fluid because tangential forces do not exist and any external forces (which must necessarily act through the mass centre of an element) also cannot give rise to a turning moment. Vorticity can arise or decay only when viscosity is present, and with an incompressible inviscid fluid it must be supposed that any vorticity present has always existed and will continue to exist unchanged.

The essential features of the mathematical scheme should now begin to emerge. For calculation, the real wing is to be replaced by a vortex system, but there are certain difficulties yet to be overcome. It is a consequence of the definitions that in an *infinite* continuous fluid, a vortex cannot begin or end, but must be continuous, like a smoke ring. It may, however, have any shape. Second, the strength of the vortex must not vary over its length, whereas in a real wing the circulation (or strength of the vortex) must decrease to zero at the wing-tips. In the ideal system used for calculation these difficulties are overcome by supposing that there are one or more line vortices situated along the span of the aerofoil—these are called *bound vortices* and always occupy the space in the fluid where the real wing is located at any instant. A particle of air which is made to rotate by the bound vortex system is swept downstream to form *trailing* or *free vortices*. The system is completed by the starting vortex (see p. 150) far behind the wing, thus satisfying the condition of continuity, but for practical purposes the trailing vortices may be supposed to stretch indefinitely downstream. The fact that in a real wing the circulation is greatest at the mid-point of the span and least at the tips is realized in the ideal scheme by superimposing a number of the simple vortices to form the sheet which ultimately rolls up into the wing-tip vortices. Prandtl's scheme is pictured in Fig. 45, and from this point onwards the calculation of the induced field (or motion around the wing) can be effected by well-known methods.

The scheme outlined above affords an excellent example of the process of idealization which occurs so frequently in mathematical physics. For the mathematician the presence of the real wing is irrelevant, except as a mechanical device which

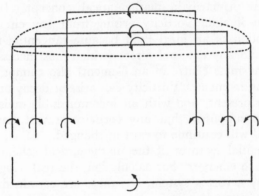

Fig. 45. The Vortex System Associated with a Finite Wing

causes circulation. For the purpose of calculation it is more convenient to deal with vortex systems, but the mathematical picture is not entirely divorced from reality. As a consequence of viscosity, the wing is enclosed in a boundary layer in which vorticity is generated, but the details of such motion are highly complicated and need not be specified for the discussion of lift. The existence of the wing-tip vortices has been amply confirmed by photographs, but acceptance of the mathematical theory is not dependent on such evidence. It is sufficient that the scheme postulated by Prandtl is possible for an inviscid fluid, and that the results are in reasonable accord with measurement.

Lanchester's discovery of induced drag is one of the milestones in the path of aerodynamics. In an inviscid fluid, a completely submerged body experiences no drag, and it seems at first sight that induced drag, which arises solely from lift and has nothing to do with viscosity, contradicts this theorem. Induced drag appears because of the discontinuities in the motion or, in more direct terms, as the vortex trail lengthens behind the wing, more and more energy passes into the wake, so that some tractive force must be present to maintain the motion. Lanchester realized that flight implied the existence of the trailing vortices and that a price had to be paid (in terms of engine power) for their existence. In flying, an engine has to do two things: (i) overcome the direct resistance of the air and (ii) keep the aircraft aloft. The trailing vortices are like railway lines laid by the aircraft in the air, and their upkeep has to be ensured.

Lanchester used the term 'aerodynamic resistance' for induced drag, and he demonstrated the fundamental difference between this and direct resistance. The direct resistance to motion increases approximately as the square of the speed (for speeds well below that of sound), whereas induced drag decreases as the square of the speed. This implies that the minimum total resistance occurs at a certain intermediate speed. In the early days of aircraft, the inefficient power units available meant that machines could do little more than reach this low speed, but modern aircraft have ample reserves of engine power and can afford to travel at higher speeds. To-day, a fast subsonic machine uses about ten per cent of its total engine power in generating lift.

The Aerofoil at High Speed

The preceding account of the action of an aerofoil is limited to incompressible flow (i.e. to speeds well below that of sound waves in air). In such conditions an infinite aerofoil moving through an inviscid fluid experiences a lift but no drag. A finite aerofoil has lift and induced drag. At high speed, comparable with that of sound, an infinite aerofoil experiences a new form of drag caused by the compressibility of the air (cf. Chapter 3).

To illustrate this, consider the simplest possible example, that of an infinite lamina moving at supersonic speed (Fig. 46). The air, on approaching the point A on the upper surface, has to turn a corner to become parallel to the lamina, which is inclined at a small angle α to the direction of the undisturbed stream. On the upper surface the air is able to do this by a smooth succession of infinitesimal expansions (Prandtl–Meyer flow, see p. 128), but on the lower surface a shock wave is formed. At the trailing edge, where the flow has to straighten out again, the position is reversed. A shock wave springs from the upper surface, with a Prandtl–Meyer expansion on the lower side.

At A, pressure on the upper surface is lowered from the atmospheric value p_0 to a value p_U; at B the value jumps back to p_0 in passing through the shock wave. On the lower surface the pressure goes through a similar series of changes in the reverse order, a sharp rise on passing through the shock wave from p_0

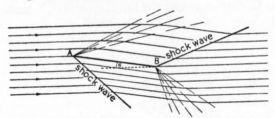

FIG. 46. Infinite Lamina at Supersonic Speed

to a value $p_L > p_0$ and finally a smooth reduction at the tail to p_0 by the Prandtl–Meyer process. It is easily shown that

$$p_U = p_0 - \frac{\rho_0 \alpha V^2}{\sqrt{(M^2 - 1)}}$$

$$p_L = p_0 + \frac{\rho_0 \alpha V^2}{\sqrt{(M^2 - 1)}}$$

where ρ_0 is the density of the air in the undisturbed state, V is the velocity of the aerofoil and M is the Mach number, or the ratio of V to the velocity of sound in the undisturbed stream. If the length of the lamina in the direction of motion is c (the chord), the force perpendicular to the lamina per unit of span is

$$F_N = c(p_L - p_U) = \frac{2\rho_0 \alpha V^2 c}{\sqrt{(M^2 - 1)}}$$

The lift is defined as the component of this force perpendicular to the direction of the undisturbed stream, or $F_N \cos \alpha$, and the drag is $F_N \sin \alpha$. Hence, since α is small ($\cos \alpha \simeq 1$, $\sin \alpha \simeq \alpha$, α in radians)

$$\text{lift} = \frac{2\rho_0 \alpha V^2 c}{\sqrt{(M^2 - 1)}}$$

$$\text{drag} = \frac{2\rho_0 \alpha^2 V^2 c}{\sqrt{(M^2 - 1)}}$$

The lift and drag coefficients are found by dividing these quantities by $\frac{1}{2}\rho_0 V^2 c$. Hence

$$C_L = \frac{4\alpha}{\sqrt{(M^2 - 1)}}$$

$$C_D = \frac{4\alpha^2}{\sqrt{(M^2 - 1)}}$$

Thus an infinite plane aerofoil, when moving at supersonic speed ($M > 1$) experiences a drag, given by the above ex-

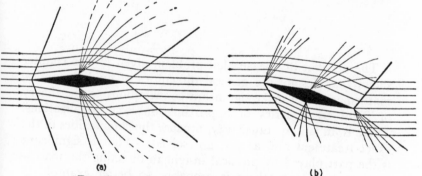

(a)   (b)

FIG. 47. Double-wedge Aerofoil in Supersonic Flow
(a) at small incidence, (b) at relatively large incidence

pressions. This drag arises entirely from the loss of energy caused by the wave-making process and is usually called *wave drag*. To this must be added the drag caused by viscosity, so that altogether an aerofoil moving at high speed experiences a high resistance.

In addition, there is a reduction in the slope of the lift coefficient–incidence curve. For a Joukowski aerofoil the rate of increase of lift coefficient with incidence is given, for small values of α, by

$$\frac{\partial C_L}{\partial \alpha} \simeq \frac{\partial}{\partial \alpha}(2\pi\alpha) = 2\pi = 6 \cdot 28 \text{ per radian}$$

At supersonic speed

$$\frac{\partial C_L}{\partial a} \simeq \frac{\partial}{\partial a}\left\{\frac{4a}{\sqrt{(M^2-1)}}\right\} = \frac{4}{\sqrt{(M^2-1)}} \begin{cases} = 3\cdot6 \text{ per radian when } \\ \quad M = 1\cdot5 \\ = 2\cdot3 \text{ per radian when } \\ \quad M = 2 \end{cases}$$

This means that, at supersonic speed, not only is the rate of increase of lift with incidence less than that found at subsonic speeds, but it decreases as the velocity increases, a fact which complicates the problem of control.

The effect of shape is marked. Fig. 47 shows the flow around a double-wedge aerofoil; it will be observed that the pattern changes sharply as the angle of incidence passes through a value equal to the semi-angle of the wedge. This is quite unlike what happens at low speeds, where change of shape has much less significance.

The account of the aerofoil problem given in this chapter shows that despite the general difficulties of fluid dynamics, mathematics can make significant contributions to the theory of flight by a series of approximations and idealizations. There is, in fact, no other way, because the complexities of the exact treatment are, as yet, unresolved. What is significant is the part played by physical insight in directing the mathematical argument; there is probably no better example in mathematical physics of the necessity of forming a clear physical picture of a process before embarking upon the analysis.

6

STATISTICS, OR
THE WEIGHING OF EVIDENCE

By ignorance we know not things necessary;
bv errour we know them falsely.

BURTON

What is Statistics?

ONE OF THE great differences between the present civilization and those of the past is that to-day it is possible for an unbiased observer to recognize trends and to predict some of their social and economic consequences with tolerable accuracy. If the science of applied statistics had been known to the Romans (together with means of collecting data and the will to remedy faults) it is possible that the Roman Empire would not have disintegrated so completely in the early centuries of the Christian era. Statistical theory is essential to good government, but the nature and purpose of applied statistics are often misunderstood. It is frequently asserted that 'anything can be proved by statistics'—a statement which is accurate only if it be taken to mean that *nothing* can be proved in this way. The twentieth-century counterpart of the facile optimism of Pippa's song seems to be a belief that things come right in the end because of the intervention of a mysterious 'law of averages'. The law of averages is mere superstition, and there is no such basis for prediction known to science.

Applied statistics is not a means of discovering new truths (such as physical laws), and, contrary to popular opinion, 'proves' nothing. The primary aims of statistical theory are: first, *to provide an objective method of testing groups of numerical data for internal consistency* and second, *to weigh the evidence for previously formulated hypotheses.* It has been pointed out (Chapter 1) that a laboratory worker spends the greater part of his time

167

in excluding unwanted effects from his experiments, so that his measurements will give him the required answer without further analysis. This presupposes a high degree of control, but there are sciences in which control is difficult or even impossible (meteorology and astronomy are good examples), and phenomena must be observed 'as they come'. In such fields there is need for a calculus to purify the data and help in the work of interpretation. This is the main task of applied statistics.

There is a second, highly specialized use for the theory: *to provide mathematical language for the enunciation and development of physical laws relating to the properties of large assemblies considered as a whole.* This application has given rise to *statistical mechanics* (in the study of molecular assemblies) and *statistical hydrodynamics* (in the study of turbulent motion).

Consider the first use defined above. What kinds of problems are treated by statistics? We can answer this most easily by giving a few examples.

1. Some property (such as the breaking strength of a specified steel), measured for a large number of specimens (say, 100) from a single lot, is found to vary from specimen to specimen. How can these results be summarized concisely and yet accurately, e.g. what single figure (average) can be quoted as representative of the lot and how widely are the values spread? These are problems of *distribution* and of *measures of dispersion*.

2. How can we define and measure the relation between two sets of variables? Can we estimate the reliability of statements such as 'Students who score high marks in mathematics also do well in modern languages'? These are problems of *correlation*.

3. When a property has to be judged by means of the examination of samples from a large lot, how should the sample be chosen to be representative of the lot, how large should it be and what change will an increase in the size of sample cause in the deductions? This is the general problem of *sampling*.

4. How can we decide whether a measured difference is real or fortuitous? For example, two groups of children

are given different diets.   In each group, some put on weight rapidly and some slowly.   With what confidence can we assert that one diet is better than another? This is the problem of *significance*.

These problems do not cover the whole field of applied statistics, but only the more commonly used sections.   There are two distinctive features.   In the first place, a number of commonplace (and therefore imprecise) notions such as 'average', 'relation', 'real difference' have to be defined unequivocally— that is, each must be given a limited but definite meaning. Secondly, rules must be derived for assigning numerical values to these concepts from an examination of the data.

Before attempting to describe the methods of statistical analysis we must pause to consider the nature of the questions and answers in more precise terms.   An experimenter seeking the accurate determination of a physical constant, such as the density of iron, will take precautions to ensure that, as far as possible, his successive measurements are exact repetitions of each other.   He will not succeed completely, but his object is to measure identical samples of a homogeneous whole in precisely the same way.   How far he can dispense with statistical theory in the reduction of his results is a very fair measure of his success in planning the experiment.   Applied statistics is best adapted to deal with nearly identical samples of a not-too-inhomogeneous whole and is thus concerned essentially with repeated observations on similar quantities or attributes. The limitations 'nearly identical', 'not-too-inhomogeneous', etc., are essential for results of real value; if the material to be analysed is very heterogeneous, the theory merely can confirm the heterogeneity.   Statistical theory is precise only in its statements about the properties of a recognizable class and is vague about the individual members of the class.   It is possible to establish with certainty the truth or falsity of the statement that the children of broken marriages are more *likely* to commit crimes than others, and such knowledge is of value to legislators and social workers, but statistics cannot help a magistrate to decide on the guilt of a solitary youth in the dock.   The statement that the velocity of the hydrogen molecule is 18,400 centimetres per second at 0° C. and 760 millimetres pressure is *statistically true*, that is, a correct deduction from experiments with large numbers of molecules, but it is not necessarily true of any individual molecule in the swarm. The statistical analysis of the behaviour of swarms of molecules leads to the gas laws, statements of high precision about gases in bulk; the same theory is more suggestive than informative about the behaviour of individual molecules. In problems of living creatures (especially human beings) the heterogeneity of any class is much greater, and the statistical deductions necessarily less precise. The popular mistrust of statistics when applied to social problems arises because of a tendency to compare statements about the behaviour of classes of human beings with our knowledge of individual men and women. The class of all men is not a man—in other words, the 'average man' does not exist—but this does not invalidate conclusions reached about the behaviour of crowds. It is as well always to bear these distinctions in mind when considering the interpretation of statistical results.

Some additional insight into the nature and limitation of the statistical method may be gained by comparison with classical mathematical physics. In most of the soluble problems of physics it is possible to assume that the various effects are simply additive—thus in the problem of external ballistics, the influences of gravity and air resistance are treated independently and expressed by separate terms in the equation of motion. If, however, this is not so—if there is interaction or cross-linking between the various effects—the system is non-linear and therefore extremely difficult to analyse by differential equations. In biology, economics, meteorology and many other subjects such cross-linking is the rule rather than the exception. One o. the chief uses of statistics is to provide a technique whereby the strands of the tangled skein can be traced to their respective origins, or, in more technical language, to decide how much of the total variation is explained by variations in a selected contributory cause. This is particularly valuable in the less exact sciences, such as biology, in which the various 'theories' seem often to be held more emotionally than rationally.

Distributions and
Measures of Dispersion

The concept of an *average* is commonplace, whether it be in relation to the weekly wages of a class of workers or the number of runs scored by a batsman. Such numbers seek to crystallize the chief properties of a group of numbers in one magnitude. Averages, however, are of more than one kind and can be formed in different ways. The most common average is what the mathematician calls the *arithmetic mean*, defined as the sum of n numbers divided by n. In many instances this quantity is admirably adapted to its primary purpose, but haphazard use of the arithmetic mean can produce results which, taken at their face value, are grossly misleading or even meaningless.

A social worker investigating a village composed of several hundred lowly-paid workers and one very wealthy man would be ill advised to base any conclusions on the arithmetic mean of the least and greatest incomes, for this figure would not be typical of any inhabitant. This is an example of the wrong kind of average. There are many simple conundrums about speed which depend on taking the average with respect to time and not distance. Here, the choice of variable is all important. The 'annual mean temperature' of a locality does not tell us if the climate is equable throughout the year or whether there are very hot summers and extremely cold winters, and it is not difficult to find many more examples of averages which, by themselves, yield no useful information. Some less naïve method of satisfying the real need for a representative magnitude must be evolved if any weight is to be attached to the result.

This requires consideration of the properties of a group as a whole. To take a simple example, the Quartermaster-General's Department has to supply uniforms for the whole Army, which contains both short and tall men. It is a simple matter of accounting to divide the recruits for a year into groups covering small ranges of height and to count the number in each group. We call this number the *frequency of occurrence* of a particular height. A graph of these numbers plotted as a continuous curve against the mid point of each

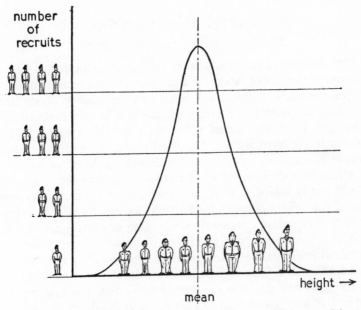

FIG. 48.   (a) Symmetrical frequency distribution (mean = mode)

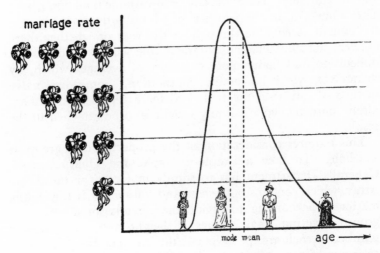

FIG. 48.   (b) Skew frequency distribution

group height-range is a curve like Fig. 48 (a). This is a *frequency-distribution diagram*. It shows immediately that there are more men of height 5½ feet (say) than any other, and also indicates the comparative rarity of very short and very tall men. As far as the Quartermaster-General is concerned, the correct 'average height' of the Army is 5½ feet, since he must supply more uniforms to fit men of this height than any other.

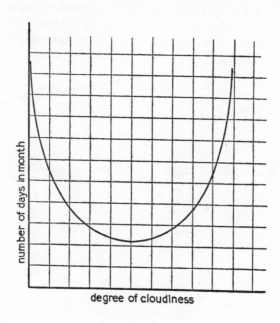

FIG. 48.   (c) U-shaped frequency distribution (idealized)

This value is not necessarily the arithmetic mean of the individual heights, but the figure about which the heights are most closely grouped. This kind of average is called the *mode* of the distribution.

The 'cocked-hat' or 'bell-shaped' distribution curve of Fig. 48 (a) is more common than any other, but there are many variations. In this example the curve is very nearly *symmetrical* (i.e. there are about as many men shorter than the mode height as there are men taller), but this is not always so. Consider, for example, the marriage rate of women in a well-defined social group.* The distribution of marriage rate (number of marriages per 1,000 women) against age is sketched in Fig. 48 (b). The rate rises steeply from 16 to just over 20 years and then falls slowly, indicating that the expectation of marriage in this group continues fairly high to nearly 40 years. A distribution of this type is called *skew*. A somewhat rare type of distribution is shown in Fig. 48 (c), which indicates a tendency for the quantity being investigated to take one or other of two widely separated values in preference to intermediate values. The classical example of this distribution is

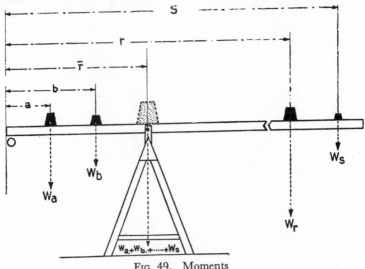

FIG. 49.　Moments

the state of the sky, which is more often completely covered with cloud or completely clear than partly covered.

The arithmetic mean, as we have seen, often fails to give a reliable representative figure because it does not give sufficient *weight* to the more important numbers. This observation shows how it can be improved, by taking a hint from elementary mechanics. A collection of bodies of weights w_a, w_b, ..., w_s, placed on a horizontal beam of negligible weight may be regarded as a system of parallel forces, for every body is attracted to the earth by gravity (Fig. 49). If we wish to

* This illustration is based on an investigation into rates of mortality and marriage among daughters of peers and heirs-apparent (Lees, *Trans. Fac. Actuaries*, 1, 276).

replace this system by one weight equal to $w_a + w_b + \ldots + w_s$, we must determine the *centroid* or *centre of gravity* of the loaded beam (i.e. the point of balance). This is done by taking *moments* about a fixed point, such as one end of the beam, O. If a weight w_r is r units distant from O, and $\bar{r}$ is the distance of the centroid from O, the ordinary theorem of moments tells us to sum the separate leverages of all the weights about O and compare this with the leverage of the total weight about the same point.   In symbols

$$\bar{r}(w_a + w_b + \ldots + w_s) = aw_a + bw_b + \ldots + sw_s$$

or

$$\bar{r} = \frac{aw_a + bw_b + \ldots + sw_s}{w_a + w_b + \ldots + w_s}$$

We call $\bar{r}$ the *weighted mean* of a, b, ..., s. If there are n weights and they are all equal to w, the expression for $\bar{r}$ becomes

$$\bar{r} = \frac{w(a + b + \ldots + s)}{nw} = \frac{a + b + \ldots + s}{n}$$

which is the arithmetic mean of the distances a, b, ..., s. Thus the arithmetic mean of a number of quantities is obtained by assuming that each quantity is as important as any other or (in colloquial but accurate language) that every term has the same *weight*.

This idea is carried over into statistics to give the general definition of the *mean of a distribution*. Suppose that f_r is the frequency corresponding to the value r (for example, f_r might be the marriage rate of women of age r). The mean $\bar{r}$ is defined to be

$$\bar{r} = \frac{af_a + bf_b + \ldots + sf_s}{f_a + f_b + \ldots f_s}$$

This definition ensures that each value of the independent variable (such as height or age) is given its proper importance in forming the mean. If the frequency diagram is a continuous curve, so that f is a function of a continuous variable x, the corresponding formula * is

$$\bar{x} = \int xf(x)dx \div \int f(x)dx$$

* The reader who is familiar with statics will recognize this as the expression for the x-coordinate of the centroid of a plane lamina. The numerator is called the 'first moment'.

In a symmetrical distribution the mean coincides with the mode. The mode is easily recognizable as the value of the variable for which the frequency is a maximum, but the mean must be calculated by the formula given above. The mean or the mode indicate the position of the distribution curve on the axis, but they give no indication of the spread of the curve. This important characteristic of a distribution is called the *measure of dispersion*.

We could give some idea of the dispersion by measuring the *range* or difference between the greatest and least values of the variable (e.g. in the example of the height of recruits, the range is about 26 inches). This is not very satisfactory because the range depends entirely on the 'fringe' values, which are usually the least reliable. A much more useful measure can be obtained by taking another hint from mechanics, this time from rotating bodies. In the motion of such a body (e.g. a flywheel) the distribution of mass about an axis through the centroid plays an extremely important part. In the dynamical equations this is conveniently expressed by the *moment of inertia*, which is found by multiplying every element of mass by the square of its distance from the axis, and summing. The moment of inertia is usually expressed as (total mass) $\times k^2$, where k is a length called the *radius of gyration*, which expresses concisely the way in which the mass of the rotating body is distributed. This concept is used in the statistical problem by the introduction of the *standard deviation* σ, analogous to k, and defined as follows. Let $a_1, b_1, \ldots$ be values of the variables *measured from the mean as origin* and $f_a, f_b, \ldots$ the corresponding frequencies. Then

$$\sigma^2 = \frac{a_1^2 f_a + b_1^2 f_b + \ldots + s_1^2 f_s}{f_a + f_b + \ldots + f_s}$$

If the curve is continuous, the expression is

$$\sigma^2 = \int x^2 f(x)\,dx \div \int f(x)\,dx$$

where x is measured from the mean.* The quantity σ^2 is called the *variance*.

---

* The reader who is acquainted with the dynamics of a rigid body will recognize this as the expression for the square of the radius of gyration of a plane lamina. The numerator is called the 'second moment'.

The standard deviation is a 'natural' unit for a frequency distribution and is much superior to the range as a measure of dispersion. If the distribution is bell-shaped and not too skew, it is generally safe to say that the whole curve will be contained within the part of the axis between $\pm 3\sigma$ from the mean. The greater part of the distribution (about 95 per cent) lies within the lines $x = \pm 2\sigma$, i.e. within two standard deviations of the mean. We can also define the asymmetry of the distribution in natural units by the formula

$$\text{skewness} = \frac{\text{difference between mean and mode}}{\text{standard deviation}}$$

Thus the three *parameters of the distribution*, mean, mode and standard deviation, tell us a great deal about the distribution in a succinct manner. The mean or the mode fix the position of the distribution on the axis, the standard deviation indicates how the data are grouped about the mean, and the three together reveal the amount of departure from symmetry. There are, of course, other ways of summarizing the main characteristics of a distribution, but these three parameters are used more than any other.

So far we have described the various frequency distribution curves only in general terms. The most famous curve of the bell variety is the Gaussian or Normal Error curve. The basic equation of the curve is

$$y = e^{-x^2}$$

where e is the base of natural logarithms. The more general expression

$$y = \frac{1}{\sigma\sqrt{(2\pi)}} e^{-x^2/2\sigma^2}$$

represents a Gaussian curve, with standard deviation σ, and the mean (= mode) at $x = 0$. This is the form of the equation generally employed in statistics.

The Gaussian curve appears very often in frequency distributions and also in applied mathematics generally. If the observed deviations from an expected value are caused by a large number of independent influences, the resulting distribution will approximate closely to the Gaussian curve. Thus, in gunnery, the fall of shot from a gun which is always laid along the same line in space follows a Gaussian distribution about the mean point of impact; this is to be attributed to the fact that the scatter is caused by random fluctuations in wind, air density, muzzle velocity, steadiness of the projectile and other effects. The concentration of smoke from a steady continuous source, measured across wind at any distance downwind of the source, also gives a Gaussian curve, because the scattering of the smoke particles is caused by random fluctuations in wind velocity and direction (i.e. by turbulence). A series of measurements of a physical constant in the laboratory usually follows a Gaussian distribution, indicating that the experimenter cannot entirely eliminate random effects.

For actuarial and other work it is usually necessary to fit a continuous curve to the observed distribution in order to smooth the irregularities in the raw data, and it is highly desirable that such curves be represented by fairly simple mathematical expressions. The fitting can be done in more than one way, but the usual method follows the systematic approach devised by Karl Pearson, who showed that a wide variety of useful curves (including the Gaussian curve) can be evolved from one differential equation, namely

$$\frac{1}{y}\frac{dy}{dx} = \frac{x + a}{b_0 + b_1 x + b_2 x^2}$$

by suitable choice of a, b_0, b_1 and b_2. Pearson's method of curve fitting consists in evaluating the 'moments' * of the distribution from the data and then forming from the moments four test quantities. From the magnitude and sign of these test numbers it is possible to choose the theoretical curve which is best fitted to represent the distribution, after which the main properties can be deduced with little further calculation. The curves are bell-shaped (symmetrical or skew), U-shaped or shaped like the letter J, so that these curves can be fitted, with varying degrees of precision, to almost every observed distribution.

* The nth moment of a particular frequency is defined as the product of the frequency and the nth power of the distance of the frequency from the value about which moments are being taken. The moment of the whole distribution is the sum of the individual moments. Thus the variance (σ^2) is the second moment about the mean. Generally, n does not exceed 4.

Correlation

During the fruit-growing season in the Pacific Coast valleys of North America a sudden severe frost at night can spell ruin, but given a few hours' warning, the farmer often can save the crop by the use of orchard heaters. Is there any simple rule which allows such conditions to be forecast? Meteorological records for these valleys show that if the difference between the

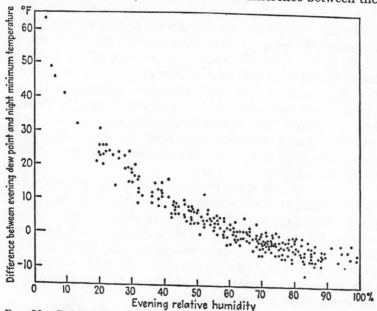

Fig. 50. Relation between night minimum temperature, evening dew-point, and evening relative humidity for a station in California. *By permission from 'Micrometeorology', by O. G. Sutton. Copyright, 1953. McGraw-Hill Book Company, Inc.*

minimum temperature of the night and the evening dew-point * is plotted against the evening relative humidity (or percentage wetness of the air), the points form a *scatter diagram* of the type shown in Fig. 50.

It is evident that there is a real connection between the variables. There are sound physical reasons for expecting a relation, because the more moisture the air contains, the more

* The dew-point is the lowest temperature to which moist air can be cooled (at constant pressure) without risk of condensation of water.

it will absorb and re-emit radiation escaping from the surface of the earth. Very dry air generally means clear skies and cold nights, shown on the diagram by a large difference between the dew-point measured at, say, sunset, and the lowest reading of the thermometer during the night. A diagram of this type greatly assists the forecaster to decide whether or not to send out a frost warning.

This is a good example of a *correlation* between two variables. The fall of temperature at night is related to the humidity of the air because dry air is more transparent to long-wave radiation from the soil and vegetation than moist air, but this is not the whole story. Cold air tends to drain into valleys from the hillsides, and wind also plays a large part, so that the relation between humidity and night minimum temperature is not precise, as is shown by the scatter of the points. No matter how accurately the meteorologist measures the dew-point and the relative humidity, he cannot expect, by the use of these data only, to predict the night minimum temperature to the same degree of accuracy.

Two measurable quantities are said to be *correlated* when certain values of one variable are more often associated with a given value of the second variable than others. Thus tall men usually, but not invariably, weigh more than short men, and heavy rainfall is more often associated with low barometric pressure than with high. What is the best way of expressing such relations mathematically and how can the closeness of the association be measured?

From Fig. 50 it is clear that one could draw freehand a smooth curve through the cluster, so that the points are about equally distributed on either side. If two variables are uniquely and perfectly related (e.g. the area of a circle and its radius) the points all fall on one smooth curve, but in problems of interest in statistics, there is always a scatter of points about a mean curve. Such a curve is known as a *line of regression* and it is clear that, in general terms, the closer the association between the two variables, the less will be the scatter about the line. The mathematical problem thus divides into two parts: (i) to find the 'best' line of regression and (ii), to discover an arithmetical function which expresses the amount of scatter about the line.

Here we shall consider only linear regressions, for which the equation is of the type

$$y = a + bx$$

If the relation between y and x were perfectly linear, any two points would suffice to determine a and b. With a scatter of points some criterion of 'best fit' must be decided in advance. There are many possible choices, the most popular being the principle of least squares which we have already met in Chapter 4, in the problem of Fourier series. This states that the best fit is obtained when the sum of the squares of the differences between the line and the actual values is a minimum. That is, having chosen the independent and dependent variables, the difference between the actual and predicted values (squared) of the *dependent* variable is minimized. This criterion, as we have already explained, ensures that there can be no accidental cancellation of large negative and positive errors, and it may be shown also that only one line satisfies the criterion, but the reader must be warned against the common fallacy that the principle of least squares necessarily gives results superior to any other method. Very often, an experienced worker can obtain a better fit by the use of a transparent scale, but this method is subjective and sometimes leads to false conclusions because of an unconscious bias on the part of the computer.

The constants a and b of the line can be determined in this way by simple but tedious arithmetic*; it is then possible to compute a 'theoretical' y for every x. This completes the first part of the problem—a line of regression has been determined— and it remains to find a mathematical function which will represent the closeness of the association between x and y. We suppose that the distribution of the values about the line of regression is approximately Gaussian and that d is the difference between the actual and the 'theoretical' values of y for every x. We have seen above that the variance (or, equally, the standard deviation) is a useful and natural measure of the dispersion in a distribution. The quantity

$$S_y{}^2 = \frac{1}{n} \Sigma d^2$$

(where n is the number of observations) gives an equally useful measure of the amount of scatter about the line of regression.

* For details see Appendix IX.

(If the distribution is approximately Gaussian, all except about five per cent of the points lie within $\pm 2S_y$ of the line.) As it stands, however, S_y suffers from the disadvantage that it depends on the units in which y is expressed, and therefore cannot be used for direct comparison with other distributions. This difficulty can be overcome by expressing S_y in terms of the natural unit of the distribution, namely the standard deviation σ_y of the y-values about their mean. The quantity S_y/σ_y is a pure number; with a perfect relation, $S_y = 0$ and for very imperfect association, S_y is indistinguishable from σ_y, so that S_y/σ_y varies between 0 and 1. In practice the number r, defined by

$$r = \pm \sqrt{\left(1 - \frac{S_y^2}{\sigma_y^2}\right)}$$

called the *coefficient of correlation*, is used to measure the degree of association between x and y. The extreme values of r are 0 and ± 1, the zero limit indicating no linear relation and unity a perfect linear relation. The negative sign indicates that the relation is inverse (y decreases as x increases) and the positive sign a direct relation (y increases as x increases).

The above analysis is probably the simplest way of introducing the idea of the correlation coefficient, but the problem can be approached more generally. We can express the relation between two variables x and y in two ways, by finding separately the dependence of y on x and of x on y. In this way we obtain two lines of regression *which coincide only when the association is perfect*. Fig. 51 shows two such diagrams, corresponding respectively to a high correlation (r about 0·9) and a very low correlation (r about 0·02) respectively. It can be shown that if $\bar{x}$ and $\bar{y}$ are the means, the equations of the lines are

$$y - \bar{y} = r\frac{\sigma_y}{\sigma_x}(x - \bar{x})$$

$$x - \bar{x} = r\frac{\sigma_x}{\sigma_y}(y - \bar{y})$$

Obviously, if $r = \pm 1$, the lines coincide and if $r = 0$ the lines are parallel to the axes, i.e. the mean value of y (or x) is the same for all x (or y), which is another way of saying that there is no linear association.

It is probably safe to say that there is no quantity in mathematics which is as much misunderstood as the coefficient of correlation. A high correlation coefficient does not necessarily mean that there is a causal relation between the two variables. There are many examples of high correlations which reflect nothing more than the fact that two series of numbers increase or decrease simultaneously, e.g. the correlation between the number of wireless licences issued in Great Britain between 1935 and 1946 and the number of mental defectives notified during the same period is 0·91*. Yet not even the most rabid opponent of

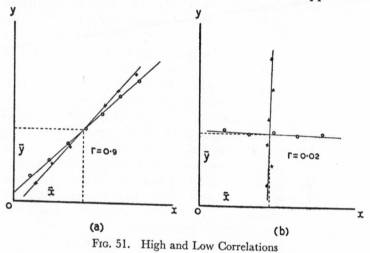

(a)   (b)

FIG. 51.   High and Low Correlations

wireless in the home would claim that either of these caused, or even influenced, the other. Here the absurdity is obvious, but there are other pairs of variables, such as the number of notifications of a certain disease and the rate of consumption of a particular food or drug, which may also yield a high correlation coefficient and yet reflect no more than a growing skill in diagnosis on the one hand and an effective advertising campaign on the other.

The lesson to be learned from these examples is that statistics is no substitute for direct experimental investigation. The observer should always establish strong physical reasons for

* This amusing example is quoted from *Introduction to Statistical Method* by B. C. Brookes and W. F. L. Dick, Heinemann, 1951.

expecting a causal relation before computing correlation coefficients. Even when there are good grounds for suspecting a causal relation, it is easy to attach too great a significance to what appears to be a high correlation coefficient. The simplest and safest way of regarding a correlation coefficient is as follows: if there is a genuine causal relation between two simply distributed variables y and x, the variance in y which can be ascribed to the influence of x is measured by r^2. To take a simple example, the meteorologist investigating night minimum temperatures might find that the correlation between the variables of Fig. 50 is as high as, say, 0·8. This means that of the observed variance in the minimum temperature, $(0·8)^2$ or 64 per cent is to be ascribed to changes in humidity. The remaining 36 per cent is not accounted for by this hypothesis. The argument can be put in a more striking fashion by considering what $r = 0·8$ implies for the accuracy of a *single* prediction. The ratio $S_y/\sigma_y = \sqrt{(1 - r^2)}$, sometimes called the *coefficient of alienation*, allows us to estimate the proportion of pure chance which remains in the estimate when the prediction has been made by the regression equation. For $r = 0·8$, the element of chance is 60 per cent. It is only when predicting *averages* that any but the very highest correlation coefficients indicate something significantly better than drawing numbers out of a hat. This explains why individual meteorological forecasts based on what appear to be high correlations frequently are badly wrong or, at the best, only a slight improvement on crude averages. The real value of the correlation coefficient lies in its power to indicate how far known external causes can be held to influence results.

Sampling

A statistical investigation is necessarily founded on the belief that the group (or *population*, to use the technical term) being investigated exhibits stable and predictable characteristics, despite the fact that the behaviour of individual elements of the population is highly variable and, in general, beyond the range of exact prediction. (This, of course, is the basis of life insurance.) Clearly, not all members of a population can be examined, and the behaviour of the whole must be deduced from a close scrutiny of a *sample*. In general, the larger the sample the greater is the chance of it being characteristic of the whole, but large samples, of themselves, are no guarantee of successful analysis. There is the well-known story of the American magazine which conducted a trial ballot before a Presidential election. The sample taken was very large—of the order of 10,000,000 voters—and yet the result was contrary to that of the real election. The names of those questioned were taken from telephone directories and registration records of cars, which meant that the sample excluded everyone not wealthy enough to rent a telephone or own a car. A much smaller sample, but chosen to include all classes of American voters in proper proportion, would have given a better prediction.

The ideal sample can be defined without difficulty—it must be unbiased and such that there is no connection between members of the sample. (Thus, for example, to test the quality of the products of a machine it would not do to take every tenth or hundredth article, because it is conceivable that the machine has a periodicity which might bias such a sample in one way or other.) We want a method by which all possible samples have the same chance of being chosen. The reader may think that to plan such a selection is not a very difficult task—by writing down numbers haphazard, for example, but actually this would be a risky procedure. Most people have 'favourite' numbers (7 is almost certain to be chosen!), and to write down a hundred such numbers without bias is nearly impossible. To overcome this difficulty statisticians have prepared tables of *random numbers*, the best known of which is the collection of 10,000 four-digit numbers made by L. H. C. Tippett. Any selection from, or combination of, these numbers or parts of these numbers forms a genuine random sequence, in the sense that there is no connection between successive members. In this way it is possible to make a *random sample* of any population. To take a specific example; suppose a manufacturer wished to ascertain the general level of quality of his products by the close examination of a sample taken from a day's output. A selection of, say, 100 items chosen by eye from stock is very likely to be biased, no matter how much the examiner tries to be impartial, but if the selection is made by means of a list of 100 random numbers sent to the store-keeper, there is a much higher probability that a representative sample will result.

We require also some means of estimating the reliability of conclusions drawn from examination of samples. If a large number of samples (each fairly large) is taken, it is found that usually the distribution of the various means approximates to the normal or Gaussian type, even when the individual population is not strictly normally distributed. Obviously, the mean of this distribution is very close to the true mean of the whole population —this expresses the popular belief that it is better to work with averages than individuals. But how close? We have already pointed out that almost the whole of a Gaussian distribution is contained within $\pm 3\sigma (\sigma = $ standard deviation) from the mean, so that having found the standard deviation for the means, we can be reasonably certain that there will be no representative sample with a mean differing by more than three standard deviations from the true mean. The standard deviation of the distribution of means calculated from samples is called the *standard error of the mean*.

If a random sample of size n is drawn from a normal population, for which the standard deviation is known to be σ, it can be shown that σ_n, the standard error of the mean, is expressed in terms of the natural unit σ by the equation

$$\frac{\sigma_n}{\sigma} = \frac{1}{\sqrt{n}}$$

Thus as n increases without limit, σ_n becomes vanishingly small. (Increasing the size of sample causes the distribution of means to become more compact and therefore more easily interpreted.) Now, if a sample of 100 items, taken at random from a much larger population with a standard deviation of 0·1, yields a mean of, say, 10, it follows that the standard error of the mean is $0·1/\sqrt{(100)} = 0·01$. This implies that it is almost certain that the true mean of the very large population lies not more than $\pm 3 \times 0·01$ from the sample mean 10, or that the true result lies between 9·97 and 10·03. This is a good example of the power of statistical theory. What has been evaluated accurately is not the true mean (this is impossible from a sample) but the residual uncertainty in the result. The theorem also indicates the advantage to be gained by increasing the size of sample. Thus, in the example given above, doubling the size of sample would decrease the range of uncertainty from ± 0.030 to about ± 0.02, and this may not be worth while in view of the labour involved in examining an additional 100 items.*

The analysis outlined above refers essentially to large samples (say, number of items exceeding 50) and may be extended to other statistical functions, e.g. for a normal population the standard error of the standard deviation is $\sigma/\sqrt{(2n)}$. If the sampling group is small, there is a tendency for the dispersion so obtained to underestimate the variation of the whole population, a feature known as *sampling bias*. It is possible to make a better estimate by the use of *Bessel's correction*. This amounts to saying that the variance must be calculated, not from the usual formula,

$$\sigma^2 = \frac{1}{n} \sum_{r=1}^{n} (x_r - \bar{x})^2$$

but from

$$\sigma^2 = \frac{1}{n-1} \sum_{r=1}^{n} (x_r - \bar{x})^2 \qquad \ldots(1)$$

or that the sample standard deviation should be corrected by the factor $\sqrt{\left(\dfrac{n}{n-1}\right)}$. For $n = 10$, this factor is 1.054, but for $n \geqslant 100$, the factor does not exceed 1.005, which is probably negligibly different from unity. Eq. (1) is said to be an *unbiased estimate* based on $n - 1$ *degrees of freedom*.

The concept of 'degrees of freedom' is extremely important in statistics. The 'freedom' refers to the number of classes for which the frequency may be given arbitrary values. In forming an average an essential condition is that the algebraic sum (i.e. the sum taking note of sign) of the residuals $x_r - \bar{x}$ must be zero, so that the calculation of σ is based on $n - 1$ independent values. To specify any distribution requires the calculation, from the samples, of one or more statistical functions and, in general, the number of degrees of freedom is the number of classes less the number of statistical functions to be determined.

---

* This calculation could apply, or example, to a precision manufacturing process with a well-established overall dispersion (σ), which is tested periodically by sampling. The result is a good check on the mean value.

The Problem of Significance

The analyses of distribution and dispersion, correlation and sampling constitute the basic tools of the statistician, and enter into almost every investigation. There remains the problem of interpretation of the results. In many of the questions put to the statistician an absolute evaluation is not sought; instead, what is required is a precise assessment of a *difference*. To see what this means, consider how we can find out if a die is 'true'. Clearly, this must mean that any face is as likely as any other to be uppermost, so that a die may be considered as a device for making a random selection from the integers 1, 2, 3, 4, 5, 6. If we make a large number of throws with an ideal die, all in the same way, we expect that every number will appear in one-sixth of the throws. Thus, if 120 throws are made, every integer should appear 20 times, but in an actual trial it would be surprising if every number appeared *exactly* 20 times—it is far more likely that some would turn up 18 times, some 19 times and so on. Provided that no number appears less than, say, 18 times or more than, say, 22 times, we feel fairly certain that both the die and the method of throwing are unbiased. On the other hand, to take an extreme example, if half the throws in a large sample gave sixes we would feel quite sure either that the die was loaded or that the throw was 'cogged'. Between these two extremes lies a region of uncertainty in which statistical theory alone can provide a reasonable method of reaching a decision.

In problems of die-throwing or coin-tossing the probability of an event can be calculated without difficulty, and if the number of trials is small (say, 10 or less) we are accustomed to finding quite marked differences from the theoretical expectation. Nevertheless, if on tossing a penny ten times we obtain ten heads or ten tails we suspect, instinctively, that something is wrong, although quite clearly such a result is *possible* even with an unbiased coin and perfectly fair tosses. Our suspicion arises from the fact that such a score is *improbable*, the mathematical odds being 500 to 1 against (i.e. the event should occur not more than once in 500 trials and might not happen at all). In the ordinary affairs of life these odds are usually accepted as decisive; no one would seriously base his conduct of affairs on such an expectation. On the other hand, although the most likely result is five heads and five tails, no one is greatly surprised at, say, seven heads and three tails. A score at least as unsymmetrical as this should occur in three out of ten trials of ten tosses each.*

These simple considerations show how more complex problems may be attacked. First we must recognize that in any situation of this type there is an unavoidable *statistical fluctuation* which arises because of the large number of random effects inherent in the event. Second, it is necessary to define some standard of comparison. In statistics this is done by formulating what is called a *null hypothesis*. In the example of the die or the penny the null hypothesis is that there is no bias, and therefore that the theoretical distribution of numbers or of heads and tails will be realized in a properly conducted trial. Third, the experimenter must fix a *level of significance*, which means deciding, in advance, the odds he regards as decisive. The criteria usually accepted by statisticians are as follows; if the experimental result, on the null hypothesis, has a probability exceeding 5 per cent (less than 20 to 1 against), the null hypothesis cannot be rejected. A probability less than 5 per cent, and certainly one less than 1 per cent (more than 100 to 1 against) means that the hypothesis is rejected.

Let us see what this means in terms of the coin-tossing trials. The null hypothesis is that the coin is unbiased. A trial gives ten heads and no tails, for which the relevant probability is 0·2 per cent. The null hypothesis may be rejected with confidence. A second trial gives nine heads and one tail, the probability of a result at least as uneven as this is about 2 per cent, and the same conclusion is reached, but with less confidence. A new coin is used, and the result is seven heads and three tails. The probability of a distribution at least as unsymmetrical as this is 34 per cent. The null hypothesis cannot be rejected, and the coin must be considered unbiased.

To apply these ideas to real problems of statistics, we must consider first that any sample gives an *estimate* of the statistical functions (mean, standard deviation) of the whole population. Suppose we have drawn a fairly large sample from a very large population and in this way have found an estimated mean (m)

* See Appendix X.

and an estimated standard deviation (s). The true values, for the whole population, are μ and σ, respectively. If the population were normally distributed and we knew σ, we could test the hypothesis that the mean is μ by writing the difference $m - \mu$ in terms of the natural unit of the sample, which is the standard deviation of the sample mean, namely, $\sigma/\sqrt{n}$. That is, we must calculate

$$(m - \mu) \div \frac{\sigma}{\sqrt{n}}$$

but since σ is unknown, we use instead the quantity

$$t = (m - \mu) \div \frac{s}{\sqrt{n}}$$

for comparison with the known properties of the normal distribution. A normal curve is completely specified by the mean and the standard deviation, and we can interpret the curve as indicating the probability that a given value will be found. (Thus the extreme values, which have a low frequency, are equally the more improbable values.) From the ideal curve we can thus assess the probability that in any one trial a measurement will deviate from the mean by a given amount. Tables have been constructed which allow this to be done at a glance.

Let us consider a specific example. A manufacturer has accepted a contract to supply, say, chemicals with not more than 10 per cent impurity. The purchaser analyses twenty-five specimens and finds that the average impurity is 12 per cent, with a standard deviation (based on 24 degrees of freedom) of 4 per cent. The statistician is asked to decide whether or not the manufacturer has met the specification. Here the null hypothesis is that he has, that the true mean amount of impurity is 10 per cent and that the difference between this and the sample mean is to be ascribed to statistical fluctuations. The statistician decides on a 5 per cent level of significance and calculates

$$t = (12 - 10) \div \frac{4}{\sqrt{25}} = 2 \cdot 5$$

From the tables he finds that the probability is about $1 \cdot 2$ per cent, that is, the odds against the difference of the means being caused by statistical fluctuations are about 80 to 1. He had previously decided not to accept odds worse than 20 to 1, and he must therefore reject the null hypothesis and inform the customer that the sample mean shows a significant difference from the specification and that the manufacturer has not met the conditions laid down in the contract.

In the above example the odds were calculated on the assumption that the test number t is normally distributed. For small samples such an assumption might lead to inaccurate conclusions. The real distribution of t was published in 1908 by the statistician W. S. Gosset, who, unlike most scientists, preferred to hide his identity under a pseudonym ('Student'). The test described above is therefore usually called *Student's t-test*. If the samples contain large numbers of items (say, 30 or above), the t-table of probabilities is nearly indistinguishable from the normal curve table.

The method described enables us to decide whether or not the sample supports the hypothesis that the mean for the whole population has the value μ. Another particularly important use for the t-test is as follows. Suppose we draw random samples of size n_1 and n_2, with means m_1 and m_2, respectively, from two similar but distinct populations. How far are we justified in asserting that any difference between m_1 and m_2 is real and not merely the effect of statistical fluctuation? Thus, for example, it might be found that 50 schoolboys from southern England had a mean height (m_1) of 50 inches with a standard deviation (s_1) of 2 inches and that 30 boys of the same age from northern England had a mean height (m_2) of 51 inches with a standard deviation (s_2) of 3 inches. Is there any justification for saying that boys from the north tend to be taller than boys from the south? The calculation is as follows:

formula

$$s^2 = \frac{(n_1 - 1)s_1{}^2 + (n_2 - 1)s_2{}^2}{n_1 + n_2 - 2}$$

(s is the combined estimate of the true standard deviation)

calculation

$$s^2 = \frac{49 \times 4 + 29 \times 9}{50 + 30 - 2} = 5 \cdot 859$$

$$t = \frac{m_1 - m_2}{s\sqrt{\left(\dfrac{1}{n_1} + \dfrac{1}{n_2}\right)}} \qquad = \frac{51 - 50}{\sqrt{(5 \cdot 859 \times 0 \cdot 0533)}} = 1 \cdot 79$$

The table of the t-function shows that this is not significant at the 5 per cent level (the t value at this level is $1 \cdot 99$). Hence the odds are a little less than 20 to 1 against the difference being caused by statistical fluctuations and most statisticians would consider this to be insufficient to justify the assertion of a real difference. This is not unexpected, because the samples are small and the spread large, but oracular pronouncements often have been made on no better basis than that given above.

A more searching criterion is given by Pearson's χ^2 (chi-squared) test, which investigates whether the results, as a whole, fit some specific law of distribution. Suppose there are n classes and that f_r is the observed frequency in the rth class. Let ϕ_r be the theoretical frequency for the specified distribution. Then

$$\chi^2 = \sum_{r=1}^{n} \frac{(f_r - \phi_r)^2}{\phi_r}$$

Obviously, the greater the difference between the observed and theoretical frequencies, the greater is χ^2. We want to know how great the difference may be without invalidating the fit to the theoretical law. The function is used with a table which indicates the probability that values of χ^2 not less than that calculated can be obtained on the specified hypothesis by random sampling and, for convenience, the probabilities are shown as certain levels, e.g. 5, 1 and 0·1 per cent.

The use of the concept can be made clear by a simple example. Suppose that certain articles, such as electrical condensers, are made by five slightly different processes and subjected to identical tests. It is found that the numbers of failures in the groups are 14, 17, 13, 16 and 25, respectively. At first sight it appears that one process is giving much poorer products than the others, but is there any real difference between them? The null hypothesis, that there is no difference, implies that there should be exactly

$$(14 + 17 + 13 + 16 + 25) \div 5 = 17$$

failures in each group. The differences from the observed frequencies are $-3, 0, -4, -1$ and 8. Hence

$$\chi^2 = \sum_{=1}^{5} \frac{(f_r - 17)^2}{17} = \frac{9}{17} + \frac{0}{17} + \frac{16}{17} + \frac{1}{17} + \frac{64}{17} = 5 \cdot 30$$

Here $n = 5$ and the theoretical distribution is defined by one condition, that the total number of failures is 85, so that there are $5 - 1 = 4$ degrees of freedom (p. 187). From the χ^2

tables, the 5 per cent value for 4 degrees of freedom is 9·49, which is considerably greater than the observed value. Hence the null hypothesis is not rejected; there is a high probability of getting a value of χ^2 greater than 5·30 and the conclusion is that there is no real evidence that any one of the five processes is better or worse than any other. However, if the numbers of failures were 14, 17, 13, 16 and 30 the value of χ^2 would be increased to 10·5, which is above the 5 per cent figure and there would be grounds for suspecting that there is a real difference between the processes.

☆   ☆   ☆

The above account has touched on only a few of the common uses of statistical theory and many more examples could be found. Perhaps the most important contribution of statistics to science is in the *planning of experiments*. The classical procedure, which is followed in most physical investigations, consists in controlling, successively, all variables except one, so that the effect of each factor is isolated and examined in turn. In biology and certain other sciences this is often impossible, or so laborious that it cannot be undertaken. Statistical theory, following lines initiated by Fisher, has found ways of overcoming this kind of difficulty (e.g. by means of the procedure known as *factorial design*) but a word of warning is necessary. Statistical theory is not a way of evading precision in thought, and the statistician is not a miracle worker who can extract results from any jumble of data. In planning an experiment, or series of experiments, which has to be analysed statistically it is wise to seek the advice of the statistician at the earliest possible stage. It is only when an investigation has been properly arranged that the full power of statistical theory can be realized.

7

MATHEMATICS
AND THE WEATHER

Who can number the clouds in wisdom?

JOB

The Problems of the Atmosphere

ONE OF THE main aims of science is to make reliable predictions. Some of these are now accepted without question; a daily newspaper, for example, publishes the times of sunrise and sunset, or announces a forthcoming eclipse, with complete confidence. Predictions of the times of high and low water and of the height of the tides at various ports are only slightly less accurate. All such predictions are quantitative and based on mathematical analysis, but another familiar prediction, the *weather forecast*, is couched in less precise language and makes no claim to high and unfailing accuracy. At first sight there seems to be no good reason for this, because mathematics is as essential in meteorology as it is in astronomy or tidal theory. The *Nautical Almanac* has been described, with justice, as 'a miracle of forecasting', and it is natural to ask why it is not possible to produce a like volume for the atmosphere. To put the matter in another way, mathematics has not yet succeeded in meteorology as it has in astronomy and it is pertinent to inquire into the reason for this.

To bring the problem into focus, consider first those parts of science where mathematics has succeeded beyond question. A little reflection shows that they fall into two well-defined groups: (i) those concerned with controlled processes from which extraneous events have been largely or wholly eliminated by careful design, and (ii) those dealing with uncontrolled natural phenomena of a highly stable kind. All physical

194

measurements in a laboratory and all operations with machines and instruments fall into the first class. Dynamical astronomy and tidal theory are typical of the second class. The solar system, for example, exhibits the most stable motion known to man, and a failure of an eclipse to occur precisely at the predicted time would indicate a catastrophic change in the system (which, of course, would force itself on our senses in other ways).

Meteorology falls into neither of these classes. Like astronomy, it is concerned with uncontrolled natural processes on a large scale, but the comparison fails in other respects. Let us try to look at the problem objectively. Held fast to the earth by the force of gravity is a vast ocean of air, a mixture of about a dozen different gases, one of which, water vapour, is able to pass very easily into the liquid and solid states (and vice versa), thereby liberating or taking up large amounts of heat. Into this mixture the sun pours energy in the form of radiation of short wavelength. Some of the radiation passes through the atmosphere to warm the land or sea, a little is absorbed by the air and much is reflected back to space from vegetation and water and by clouds. The heated earth, in turn, gives out radiation of long wavelength (infra red), some of which is absorbed and re-emitted by water vapour and clouds and some is lost to space. (In technical terms, the balance sheet of radiant energy shows changes, not only with the time of day, season and locality, but also with the composition of the atmosphere and with the distribution of land and water over the globe.) This, however, is not all; the unequal heating sets up currents in the atmosphere which circulate quantities of air from one location to another, in the course of which water is evaporated or condensed in vast amounts. Almost every element of the system reacts significantly with every other and cross-linking, the bugbear of the mathematician, is the rule rather than the exception. We can sum it up by saying that the atmosphere is an incredibly complicated steam engine, with the sun as furnace, located on a spinning globe of highly irregular surface. The remarkable thing is not that mathematical prediction has failed, but that any form of forecast of the details of such a complicated system can be contemplated, let alone expected to produce reasonable results.

The intricate nature of the meteorological problem is perhaps better realized by a comparison, not with astronomy, but with economics. Both subjects deal fundamentally with problems of energy transformations and distributions—in economics, labour produces goods and other forms of wealth and meteorology traces the distribution of energy from the sun. Both systems are affected by complicated and seemingly capricious external influences. Economics has to reckon with man in his infinite variety, meteorology with effects arising from the irregular character of the surface of the earth and the uneven distribution of water vapour in the atmosphere. We are painfully aware that there is no universally accepted theory in economics and very little hope of establishing such a theory in the present generation. In plain language, economics is too difficult to be developed quickly into an exact science, and much of the political strife of to-day arises from the refusal of enthusiasts to admit this simple truth. We may go further than this. In problems of economic behaviour variations in human psychology introduce disturbances of an apparently irrational (and therefore unpredictable) kind into any planned system, and for this reason it is doubtful if a genuine mathematical description of the whole economic field, with the implication of exact prediction, is possible. Much the same arguments may be applied to meteorology. To bring the weather within the scope of exact mathematical analysis demands the formation of something approaching a universal theory of atmospheric motions. It is only rarely that universal theories appear even in highly developed subjects like physics, and research, for the most part, is directed towards the exploration of limited fields as they become ripe for exploitation. Quite apart from this aspect, however, there arises the possibility (as in economics) of extraneous random influences. We do not yet know how far unpredictable elements such as sunspots or bursts of cosmic radiation may affect atmospheric motions; if they play a significant part, weather is subject to an inherent indeterminacy which makes real long-range forecasting impossible.

The solution of the tidal problem is facilitated by well-marked periodicities, but there are no such regular oscillations in the atmosphere, and to-day, most meteorologists have abandoned the search for such aids. Nor, despite much labour, has it been found possible to enunciate simple causal relations which connect the states of the atmosphere at different times. The purely statistical approach to the problem is now discredited by all except a few enthusiasts, and to-day attention is concentrated more on the direct line of attack. The atmosphere is subject to the same laws as any other physical system; the difficulty arises mainly from the fact that the laws themselves are simply generalities and we need precise knowledge (in the form of initial and boundary conditions) before we can make effective use of them. Such knowledge is hard to come by, especially when it is considered that for large areas of the globe (such as parts of the oceans, deserts and polar regions) there are no meteorological records. Even if such knowledge could be obtained, however, there are other serious obstacles to progress. The equations of motion, being non-linear, are extremely difficult to handle and speed clearly is an essential factor in any realistic system of weather forecasting. Electronic calculating machines seem to be essential if the computation is to 'keep up with the weather', and for this the task of codifying and feeding in the masses of data at the required rate presents almost insuperable difficulties at the present time.

However, this is looking ahead too far—many problems can hardly be stated at the present time, let alone solved. The only possible approach is by way of simple, highly artificial problems which admit of approximate solutions. Such solutions, when compared with real situations, may furnish important clues to the next step. Because of the great complexity of the work we can discuss here the setting and solution of the simpler problems only, and for this we must examine first certain idealized states of the atmosphere.

The Atmosphere at Rest

The simplest model for the atmosphere is a stagnant ocean of air resting on a horizontal plane and extending indefinitely upwards. In this system pressure (p), density (ρ) and temperature (T, measured on the absolute scale) depend only on height (z). The pressure of the atmosphere, the quantity measured by a barometer, is simply the weight of the column of air of unit cross-section area above the barometer. Consider a slice of the column of infinitesimal thickness δz. The weight of the slice, $g\rho\delta z$ (g = acceleration due to gravity) must equal the difference in pressure between the top and bottom of the slice, which we call $-\delta p$. Then

$$\delta p = -g\rho\delta z$$

and in the limit, as $\delta z \to 0$, we obtain the differentia equation

$$\frac{dp}{dz} = -g\rho$$

This is the fundamental relation of atmospheric statics, called the *hydrostatic equation*.

This simple relation indicates how the barometer is used to measure height. Density is never measured directly, but by making use of the *equation of state* of a perfect gas, namely

$$p = R\rho T$$

where R is the gas constant, the hydrostatic equation can be rewritten in terms of pressure and temperature, both of which are easily measured. The equation becomes

$$\frac{dp}{p} = -\frac{g}{R}\frac{dz}{T}$$

We cannot solve this equation unless we know how temperature varies with height. If we replace T by its average value (T_m) up to the height z, the solution is

$$p = p_s e^{-gz/RT_m}$$

where p_s is the pressure at the surface ($z = 0$). Thus pressure decreases exponentially with height in an atmosphere of constant temperature. Also, if we know p_s, and can measure or estimate T_m, we can use this relation to find approximate heights from pressure readings. The simplest form of altimeter is an aneroid barometer with a dial engraved in feet or metres instead of units of pressure and arranged so that it can be set for different surface pressures.

The rate of decrease of pressure with height in the real atmosphere is sufficiently regular to justify the use of this kind of approximation (especially if allowance is made for changes in temperature) in nearly all conditions. Temperature by itself varies much too irregularly with height to be used for the same purpose. On the whole, temperature decreases with height from the surface up to a level known as the *tropopause* (about 30,000 feet over the British Isles), and then remains fairly constant, in the *stratosphere*, but below the tropopause it is quite common for temperature to increase with height in intermediate layers.

In exactly the same way it can be shown that density decreases exponentially with height in an atmosphere of constant temperature. In this case surfaces of equal pressure and of equal density coincide. A stratification of this kind is called *barotropic*. In general, motion destroys a barotropic state, and the real atmosphere is nearly always *baroclinic* (i.e. surfaces of equal density intersect surfaces of equal pressure). In a barotropic atmosphere pressure is a function of density only and is unaffected by temperature changes and variations in water-vapour content. The assumption of barotropy thus implies a considerable idealization, similar to the 'ideal fluid' of hydrodynamics (Chapter 5) which, being homogeneous and incompressible, obviously is barotropic.

In weather analysis, heights are rarely expressed in units of length (feet or metres), except in the final forecast. The natural way of specifying levels in the atmosphere is by pressure, and meteorologists are accustomed to think of conditions in the upper atmosphere as taking place at say, the 700 or 500 millibar * levels rather than at so many thousands of feet above the surface. The actual height of an isobaric (equal pressure) surface naturally depends on the surface pressure and therefore varies with the prevailing weather conditions.

We must now consider how motion can be introduced into an idealized atmosphere.

Fluid Motion and Richardson's Dream

The difficulties of solving the equations of fluid motion, to which reference has been made earlier in this book, are especially severe in meteorology. In aerodynamics and in many

---

* Professional meteorology has long abandoned length (of a column of mercury) as a measure of pressure. A millibar (mb) is defined as a thousand dynes per square centimetre, the dyne being the unit of force in the metric system. Sea-level pressures are of the order of 1000 mb.

other applications of fluid dynamics it is often possible to simplify matters by the process known as *linearization*, but this always introduces restrictive initial assumptions. Two-dimensional irrotational motion, which allows the complex variable to be used so effectively, is a typical example (see Chapter 5). In other branches of applied mathematics approximations made by ignoring squares and products of the velocities are tolerable because the motion is always very slow, e.g. in the investigation of the fall of a speck of dust or a mist particle in a gravitational field. Similar assumptions are made in tidal theory.

If u, v and w are the x, y and z components of velocity, respectively, of an incompressible viscous fluid of density ρ and viscosity μ, the typical equation of motion is

$$\rho \left(\frac{\partial u}{\partial t} + u \frac{\partial u}{\partial x} + v \frac{\partial u}{\partial y} + w \frac{\partial u}{\partial z} \right) = \mu \left(\frac{\partial^2 u}{\partial x^2} + \frac{\partial^2 u}{\partial y^2} + \frac{\partial^2 u}{\partial z^2} \right) - \frac{\partial p}{\partial x} + X$$

$$\underbrace{\qquad\qquad}_{\text{inertia terms}} \qquad \underbrace{\qquad\qquad}_{\text{viscosity terms}} \quad \underset{\text{gradient}}{\underset{\text{pressure}}{}} \ \underset{\text{term}}{\underset{\text{forces}}{}}$$

where X represents the x-component of external forces, such as gravity or those arising from the rotation of the earth. There are similar relations for v and w, and the whole set is known as the Navier–Stokes equations (p. 65). These equations, together with the equation of continuity (which expresses the conservation of matter), in theory determine velocity and pressure as functions of position and time when the initial and boundary conditions of a particular problem are properly specified. Unfortunately, no mathematical technique has been discovered which enables exact solutions to be found, and in the majority of applications the equations are intractable.

The Navier–Stokes equations are simply the expression of Newton's second law of motion applied to fluids. The second law states that the product of the acceleration of a particle and its mass is equal to the sum of the forces acting on the particle. In fluid dynamics density represents mass, and the terms inside the bracket on the left-hand side of the equation represent acceleration. There are four such terms because in an infinitesimal interval of time a fluid particle may change its velocity because the field is changing $(\partial u/\partial t)$ and also because it moves to a new position in the field $(u \partial u/\partial x$ etc.$)$. The quantities on the left-hand side are called the *inertia terms*. Their general effect is to accentuate any differences of velocity as time proceeds.

The *viscosity terms* (those multiplied by μ) have the opposite effect. They involve the Laplacian (p. 66) and arise from the internal friction of the fluid. These terms express the action of the incessant molecular agitation in smoothing away sharp changes in velocity by the process known as *diffusion*.

The term $\partial p/\partial x$ is the x-component of the *pressure gradient*. In the atmosphere the horizontal distribution of pressure is very irregular, being high in some places and low in others. In meteorology such areas are called *anticyclones, troughs, depressions* and the like, and the whole field changes continually. A fluid, in general, moves from regions of high to regions of low pressure, but in the atmosphere the rotation of the earth deflects the motion so that winds do not blow directly into a centre of low pressure but around it. The *forces terms* express this effect, which, however, involves the velocities, so that the unknowns of the equation appear also in the term denoted by X.

The Navier–Stokes equations indicate that fluid motion conforms to a complicated balance, with the pressure gradient as the driving force, in which the tendency of the inertia terms to steepen velocity gradients is directly counteracted by the diffusing action of viscosity, the whole being subject to modification by the external forces. The mathematical difficulties arise mainly from the non-linearity of the inertia terms. These terms contain squares and products of the unknown velocities, a circumstance which forbids the mathematician to use methods which are effective in the solution of linear equations. In particular, it is no longer possible to build up the complete solution by adding together simple solutions (this is the principle of superposition referred to in Chapter 4). In plain language, the system is strongly cross-linked and the various effects are not simply additive. In meteorology it is not possible to remove the square and product terms *ab initio* because the motion is not small. A further complication, but less serious, is that the equations, as given above, apply only to an incompressible fluid. Compressibility appears in fluid motion chiefly as a result of high speed (Chapter 5), but the strong tendency to vertical motion and the great depth of the atmosphere mean that density changes must be considered in the meteorological problem, although winds never approach the speed of sound. The inclusion of density variations means that the equations of thermodynamics must be brought into the complete problem.

If we could specify the initial and boundary conditions and afterwards integrate completely the equations of motion, we should be well advanced towards the goal of mathematical forecasting. The earliest attempt to do this was made in 1922 by the English meteorologist L. F. Richardson, who published an account of his studies in a book entitled *Weather Forecasting by Numerical Process*. Richardson's attempt was premature in that at the time there were far too few observations for his purpose, especially in the upper air, and his one example of a truly mathematical forecast was a complete failure. The failure, however, was of no significance; what mattered was that Richardson indicated what is perhaps the only feasible line of attack on this formidable problem. His main concept was that the equations should be solved by step-by-step numerical methods, very similar to those used in ballistics (Chapter 3), but at the time he wrote his book the electronic computer had not been thought of and Richardson realized that to 'keep up with the weather' would require a veritable army of computers working together like a monstrous orchestra under the 'baton' of a directing mathematician. To-day the network of observations is more complete and electronic machines work at incredible speed, but we are still a long way from realizing Richardson's dream.

Richardson, in his pioneer work, tried to solve the whole problem. Present-day investigations (such as those in the Meteorological Office and in the United States Weather Bureau) are less ambitious and the approach is by the familiar road of simplified models leading to more complete studies. The starting point for most of this work is a remarkable approximation known as the *geostrophic balance*, which in many ways is the fundamental theorem of dynamic meteorology.

The Geostrophic Balance

Except in thunderstorms or in the neighbourhood of mountains, the vertical component of the velocity of the air (w) is much less than the horizontal components (u, v). As a first step toward a workable model we may suppose that the wind is always horizontal, so that one of the Navier–Stokes equations, and all terms involving w in the others, vanish. Next, we may take a hint from Prandtl's boundary layer theory (p. 140), which says that friction exercises a first-order effect only near a rigid boundary, like the surface of the earth. If we confine attention to motion at heights not less than about 1500 feet the removal of friction terms from the equations does not make the model unrealistic. This step has the important consequence that the equations are reduced in order, for all the second-order derivatives are multiplied by μ. As an additional simplification we may suppose that conditions are steady, or that the motion at any point is independent of time, which removes the terms $\partial u/\partial t$ and $\partial v/\partial t$ from the equations. In the resulting system the equations state that the horizontal pressure gradient and the deviating force caused by the rotation of the earth balance the product of the density and acceleration of a moving fluid particle.

The precise effect of the rotation of the earth must now be considered. It can be proved that if the earth were perfectly smooth a particle on its surface would not remain in the same position indefinitely, but would slide toward the equator. A force, depending on the latitude (or distance north or south of the equator) and on the angular velocity of the earth about its axis is required to keep the particle at rest. If ω is the velocity of rotation of the earth (about 7×10^{-5} sec^{-1}) and ϕ is latitude, an element of air, because of the rotation, is subject to a force $2\omega\rho \sin \phi \cdot V$ per unit mass, where V is the resultant velocity. This is usually called the *Coriolis force*. Its effect is to produce an acceleration of magnitude $2\omega \sin \phi$ times the velocity of the particle, directed at right angles to the motion, which leaves the speed of the particle unchanged, but alters its direction.

The Coriolis force acts on all bodies moving in the atmosphere and thus affects the flight of projectiles. The effect is negligible on, say, the trajectory of a golf ball, but can be detected in gunnery and must be allowed for in long-range bombardments. It is particularly important in meteorology and supplies an essential element in the geostrophic approximation.

The reader is probably familiar with the general form of the 'weather map' which is the basis of forecasting. The most prominent features of such a map are the *isobars*, or lines

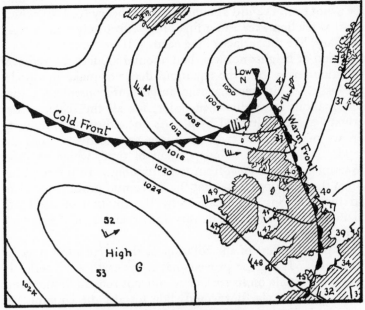

Fig. 52.   Weather Map, showing Isobars and Fronts

joining places having the same pressure at the time of observation. The isobars usually follow well-defined and familiar patterns, such as that shown in Fig. 52, which illustrates the passage of a depression, or region of low pressure, over the British Isles. The winds, shown by arrows, blow around the centre, following the invariable rule known as *Buys Ballot's law*, that in the Northern Hemisphere an observer standing with his back to the wind has low pressure on his left hand. (The rule is reversed in the Southern Hemisphere.) An isobaric chart represents a section of the pressure field at a given instant, and the gradient of pressure over any locality is indicated by the spacing of the isobars. Thus we may consider that the terms $\partial p/\partial x$ and $\partial p/\partial y$, the components of the pressure gradient, and the factor $2\omega\rho \sin \phi$ in the Coriolis term are known to the meteorologist.

In the geostrophic approximation it is assumed that an element of air moving in a known pressure field is subject only to the pressure gradient and the Coriolis force. The *geostrophic wind* is a motion along the isobars which produces a deviating force to balance the pressure gradient. If we call this wind G we have

$$G = \frac{\text{pressure gradient}}{2\omega\rho \sin \phi}$$

In this equation the right-hand side consists of known quantities, so that the magnitude of G may be found without difficulty from the isobaric chart.

The geostrophic balance is of great importance in theoretical meteorology, because it represents the simplest type of atmospheric motion with some claims to reality. It is possible to obtain a somewhat better approximation by taking into account the effect of the curvature of the isobars. Near the centre of a depression the isobars approximate to small circles, so that air moving around them is subject to a strong centripetal acceleration. In this extended system the factors which must balance are:

(i)   the pressure gradient, producing a force perpendicular to the isobars, directed from high to low pressure,

(ii) the Coriolis force, producing an acceleration towards the right (in the Northern Hemisphere) perpendicular to the direction of motion,

(iii) the centripetal force, perpendicular to the direction of motion, producing an acceleration of magnitude V^2/r, where r is the radius of curvature of the isobars.

The balance is expressed by the equation

$$2\omega \sin \phi(V - G) = \pm V^2/r$$

the plus sign being for anticyclones and the minus sign for depressions. V is called the *gradient wind*; it will be observed that the gradient wind approaches the geostrophic wind for large values of r (i.e. for straight isobars).

A detailed comparison between gradient winds calculated from isobaric charts and winds measured by pilot balloons at 1500 feet was made many years ago by the British meteorologist E. Gold, who found that at this height the wind blows almost exactly along the isobars but with a speed slightly less than the gradient wind. In routine forecasting the geostrophic wind is evaluated by means of a graduated transparent scale placed at right angles to the isobars at the place for which the wind velocity is required. The scale is moved until the graduations agree with the spacing of the isobars, so that in effect the pressure gradient is measured and automatically divided by the fixed quantity $2\omega\rho \sin \phi$. For most purposes the geostrophic wind is a sufficiently good approximation; the gradient wind is more troublesome to assess because of the difficulty of estimating the radius of curvature of the path of the air particles.

The gradient wind is thus determined by a simple quadratic equation. An analysis of the properties of roots of this equation is of interest in that it shows clearly how the mathematician is enabled to reject some mathematical consequences of his model because they are inappropriate physically. One solution (of either the anticyclone or depression equation) is found to represent a condition in which the direction of circulation is opposite to that of the earth. Such a system is consistent with the original equation, but is rejected because there is no mechanism in the atmosphere which could produce *large-scale* rotations of this type, and such systems have never been observed. The model, in fact, is too general. This is obvious in the simple analysis of the gradient wind, but in more complicated models it is not so easy to decide whether a motion is physically possible. Another interesting, and valid, deduction from the model is that winds in an anticyclone cannot exceed a certain strength, whereas for a depression there is no limit to the wind force, as far as the mathematician is concerned. It is a matter of common observation that anticyclones are usually associated with low winds.

The geostrophic model has many points of close contact with the real atmosphere, but it is necessary to point out that the analysis outlined above has no direct bearing on the problem of forecasting. In middle latitudes the geostrophic wind is a first approximation (a very good one) to the actual wind, and this knowledge is invaluable to the forecaster, but the discussion throws no light on the development of the pressure and motion fields with time. In fact, because the balance is maintained without vertical motion, there can be no sequence of weather in a geostrophic system. In this sense the geostrophic balance must be regarded as an excessive idealization of the rea state of affairs, so that departures from the balance are of the greatest importance.

Routine Weather Forecasting

In order to form a reasonable estimate of the possibilities of mathematical forecasts we need to know something about the methods which are used by 'official' forecasters to-day. The basic information (facts about present weather) comes from a network of meteorological stations scattered throughout the settled areas of the world, and from ships on the high seas. At the present time it is probably fair to say that the meteorologist has all the information that he can usefully digest for surface conditions and rather less than he needs for the upper air. The forecaster plots these observations on large maps covering, say, western Europe and the adjoining seas, and his first step is usually to draw lines of equal pressure (isobars) for the surface. In addition, he transfers the results of soundings by radio-sondes * and aircraft to thermo-dynamic diagrams. His information is displayed in selected two-dimensional 'snapshots' of the atmosphere, together with a few vertical views. In drawing the isobars the forecaster is influenced, not only by maps for the previous period, but also by what is called 'air-mass analysis', a procedure based on the fact that streams of air of different origin tend to preserve certain properties as they move. The junctions of such air masses usually appear as lines of discontinuity, called 'fronts', in the pressure field (see Fig. 52).

The normal process of short-range weather forecasting (say, for periods of six, twelve or twenty-four hours ahead) consists

---

* Balloons carrying equipment which automatically transmits values of pressure, temperature and humidity to observers below.

chiefly of an attempt to foresee changes in the pressure field by a combination of past experience and physical reasoning. The actual weather is deduced as a consequence of the movements of the field. Experience has shown that more often than not definite kinds of weather accompany certain familiar patterns of isobars, e.g. the stormy, unsettled weather of a depression and the fine warm weather of a summer anticyclone. Even so, the connection between the pressure distribution and the weather is far from unique, and the above relations are only statistically true. (This is the reason why disgruntled owners of barometers marked with weather legends sometimes declare that their instruments are 'inaccurate'.)

Weather prediction, at the present time, is more of an art than a science. There are very few quantitative rules, and calculation, in the strict mathematical sense, is rarely used. It is probably safe to say that no two forecasters follow precisely the same method or reason in exactly the same way, but the broad basis of short-range forecasting is much the same everywhere. Although the drawing of isobaric charts and, especially, the insertion of fronts, are to some extent subjective, there are certain outstanding features of all pressure patterns. The forecaster concentrates attention on these and argues future developments by a process which resembles extrapolation in mathematics. (Attempts have been made to produce a systematic classification of pressure systems, but because of the many variables and enormous range of possibilities no two weather maps are ever exactly alike, and forecasts which simply repeat the exact sequence of weather observed in the past in like systems have not been very successful.) The extrapolation is based on the recognition of certain tendencies which are expected to control the development of the system during the next six, twelve and perhaps twenty-four hours. These tendencies are summarized in the preamble to the published forecast ('A depression on the Atlantic is approaching the British Isles'), and the sequence of weather predicted for the ensuing period is that which the experience of the forecaster leads him to associate with, say, the passage of a region of low pressure over these islands. The value of this method clearly depends markedly on the range of the forecast. For six hours ahead such predictions are generally reliable, but the degree of success falls rapidly as the period lengthens and, in general, straightforward extrapolation is of little or no use for periods of the order of a week except on occasions when the pressure field is very stable.

The essential requirements for success in forecasting by this method are a liking for the job, experience and physical insight, qualities not unlike those which go to make a good general practitioner in medicine. Even short-period fore-casting is highly subjective, and a profound knowledge of the deeper aspects of meteorological research is no guarantee of success in this difficult art. The quality which is shared by all good forecasters is an intense interest in the weather which leads them, in time, to develop a 'nose' for coming changes, that is, a power of reasoning which can hardly be expressed in precise physical language and not at all in terms of mathe-matics.

The main aim of meteorological research is to provide aids which will reduce the subjective element in forecasting to acceptably small proportions. How far this can be done is still very much in doubt, but we can be sure that the need for individual skill in forecasting will never disappear. There are good reasons for believing that the exact sequence of weather at any given locality can never be calculated for long periods ahead, in the way, for example, that the astronomer can pre-dict the configurations of the heavenly bodies. We shall now look at the evidence for this assertion, and see how mathematics can assist in the future development of meteorology.

Mathematics and Forecasting

Before we examine the possibilities of improving weather prediction by mathematics, we must inquire into the meaning of the phrase 'accuracy of forecasts'. In most branches of physical science the agreement between theory and observa-tion is tested, ultimately, by comparisons between numbers. This is not so for weather prediction, because forecasts tend to be qualitative rather than quantitative. Even predictions of temperature, which are usually numerical, cannot always be checked (e.g. the statement that 'there will be frost in sheltered places' may be true of some, but not all, localities or there may be no records for such places to decide the question). The human element also enters—a forecaster who tries to be precise in his statements is more easily proved wrong than one who 'hedges' by more general statements. It is a matter of great difficulty to frame an objective method of testing the accuracy of forecasts which is fair both to the meteorologist and the user.

At the present stage, no mathematical method tries to predict all the elements of the weather. The aim of the mathematical theorists is to find a basis for the calculation of future pressure or motion fields from existing fields. The success or failure of such schemes can be decided only by comparison of the forecast and actual maps on a visual basis. A numerical assessment of success is virtually impossible—if the computations produce a distribution which reproduces the main features of the real field, the mathematical theory may be said to have achieved its purpose. Beyond this it is not safe to go at present.

The earliest attempts to bring mathematics into large-scale atmospheric processes were by means of what we may call 'mechanical' models. The typical depression, with air moving around a centre of low pressure, resembles the classical vortex (p. 160) and thus it was natural to consider first the dynamics of a revolving horizontal disc of incompressible fluid when fluid is withdrawn from the centre by some unspecified process. The resulting motion is equivalent to that found by combining a simple line vortex of the type discussed on p. 160 with a solid rotation. Such a model exhibits many features found in tropical cyclones (tornadoes), but is far too simple to represent adequately the depressions of our latitudes. Another possible model is the so-called 'cartwheel depression', a disc of fluid spinning about a vertical axis and drifting horizontally. This again has points of contact with reality.

Models of the above types are now of academic interest only. The modern approach begins with the so-called *wave theory of cyclones*, evolved by Norwegian meteorologists under the leadership of V. Bjerknes. In many branches of applied mathematics results of great value have been obtained by what is called *perturbation theory*. In this, a simple basic motion (or state of rest) is postulated and the effects to be studied, compared with the system as a whole, are regarded as disturbances which are 'small' in the mathematical sense. Squares and products of the additional velocities, when considered on the scale of the whole motion, can be neglected, so that the hydro-dynamic equations become linear. In this way Bjerknes was led to the idea that extra-tropical cyclones (the depressions of our latitudes) begin as wave-like perturbations in the *polar front*, the surface of separation between the warm westerly currents of the middle latitudes and the cold easterly currents from the polar regions. This concept gave forecasting its greatest impetus, by introducing the ideas of 'air masses' and 'fronts', but in the subsequent development of the theory in the forecaster's hands the mathematical aspects have tended to be ignored. To-day, the theory has become so involved that the original impetus may be considered to have spent itself. It is the familiar story of an over-simplified scheme, for rela-tively few disturbances conform exactly to the polar-front scheme, but the perturbation equations remain as a subject for serious study.

Modern forecasting techniques concentrate attention on un-dulations in surfaces of equal pressure, rather than on varia-tions of pressure on surfaces of equal height, and this lead is followed in the mathematical approach. Consider, for ex-ample, the 1000-mb and 500-mb surfaces. The former is roughly at sea-level and the latter lies about 18,000 feet higher (in latitude 45°, approximately mid-way between sea-level and the tropopause). As the weather changes with the passage of depressions and anticyclones, the heights of these surfaces above a fixed datum alter from point to point. Such variations may be specified in two ways: by *contour-height* above mean sea-level and by *thickness*, or the distance between two isobaric surfaces. Contour-height thus expresses the absolute topography of an isobaric surface and thickness, the relative topography of any two surfaces.

The contours (lines of equal height) of an isobaric surface are very similar to isobars, and winds blow around the 'lows' and 'highs' of the surface in the senses indicated by Buys Ballot's law. To interpret thickness we appeal to the barometric-height formula (p. 198). This may be written

$$z - z_0 = -T_m(g/R) \log_e(p/p_0)$$

where $z - z_o$ is the difference in height (thickness) of the isobaric surfaces defined by p_o and p, and T_m is the mean temperature of the stratum between the surfaces. Thus the thickness of a layer is proportional to its mean temperature, and lines of

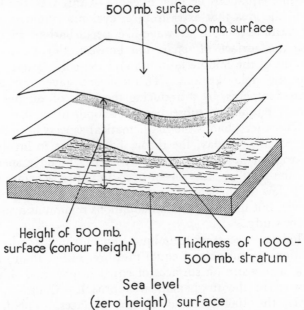

FIG. 53.   Isobaric Surfaces in the Atmosphere

equal thickness are effectively lines of equal mean temperature (isotherms). Charts of thickness have 'cold' and 'warm' centres like the 'lows' and 'highs' of contour charts.* The component of the wind that blows around these centres is called the *thermal wind*. Mathematically, this is the vector difference between the geostrophic winds on the two specified isobaric surfaces. The thermal wind is related to the thickness lines in much the same way that the geostrophic wind is related to the isobars, blowing cyclonically around a cold centre and anticyclonically around a warm centre.

* Contour and thickness charts are published in the *Daily Aerological Record* of the Meteorological Office; isobaric charts are given in the *Daily Weather Report* and by the Press.

The first step towards a mathematical system of forecasting consists in an idealization of atmospheric motions by the removal of the vertical component. Contour-height h is brought in to express the effects of pressure gradient by the use of the hydrostatic equation (p. 198). As a result, density and pressure do not appear explicitly in the transformed equations, in which the contour-height h is expressed in geodynamic measure. (For practical purposes, geodynamic and ordinary measures are much the same.) The equations of motion of an inviscid fluid on the rotating earth are then:

$$\left. \begin{aligned} \frac{\partial u}{\partial t} + u\frac{\partial u}{\partial x} + v\frac{\partial u}{\partial y} &= \lambda v - \frac{\partial h}{\partial x} \\ \frac{\partial v}{\partial t} + u\frac{\partial v}{\partial x} + v\frac{\partial v}{\partial y} &= -\lambda u - \frac{\partial h}{\partial y} \end{aligned} \right\} \quad \ldots(1)$$

In these equations u and v are component velocities along horizontal axes Ox, Oy, pointing east and north, respectively, and λ is the Coriolis parameter $2\omega\sin\phi$. The geostrophic balance is expressed by making the left-hand sides of the equations vanish, so that if u_g and v_g are the components of the geostrophic wind, we have

$$u_g = -\frac{1}{\lambda}\frac{\partial h}{\partial y}, \ v_g = \frac{1}{\lambda}\frac{\partial h}{\partial x}$$

The purpose of the investigation is to study changes in h, which is really the same thing as studying perturbations of the pressure field.

The idealization described above is justified *a priori* because the large weather-producing disturbances are characterized by small vertical motions and almost-geostrophic winds, with the hydrostatic equation very nearly satisfied. The course of the analysis is suggested by the principle of the *conservation of vorticity*. The vertical component of vorticity, ζ has been defined on p. 160 by the equation

$$\zeta = \frac{\partial v}{\partial x} - \frac{\partial u}{\partial y}$$

In meteorology ζ is called the relative vorticity. When the spin of the earth is added we have the *absolute vorticity* $\zeta + \lambda$, denoted by the symbol η, and this is the quantity which is significant in the atmospheric problem.   We also introduce the symbol D/Dt for $\partial/\partial t + u\partial/\partial x + v\partial/\partial y$; this operation is called 'differentiation following the motion of the fluid'.   To bring η into the equations we differentiate the first equation of the set (1) with respect to y, the second equation with respect to x, and subtract.   The result is

$$\frac{D\eta}{Dt} = -\eta\left(\frac{\partial u}{\partial x} + \frac{\partial v}{\partial y}\right) \qquad \ldots(2)$$

Equation (2), known as the *vorticity equation*, is one of the basic relations of dynamical meteorology.   The quantity $\partial u/\partial x + \partial v/\partial y$ is the two-dimensional form of the divergence of the motion (see p. 54).   If the winds are strictly geostrophic and the variation of λ with latitude is disregarded, it is easily shown that the horizontal divergence of the wind vanishes. Thus

$$\frac{D\eta_g}{Dt} = 0 \qquad \ldots(3)$$

where η_g is the absolute vorticity of the geostrophic motion. This result is known as the *equation of conservation of absolute vorticity*.   In a geostrophic system the motion and pressure fields always adjust themselves to keep the absolute vorticity of the system unaltered.   It is also easily seen that, if λ be regarded as a constant,

$$\eta_g = \frac{1}{\lambda}\nabla^2(h) + \lambda \qquad \ldots(4)$$

where ∇^2 is the Laplacian (see p. 66).

We have now all the equations required to formulate a simple model for forecasting purposes.   The tendency to conserve vorticity, or spin, is often seen in real fluids, for example, in eddies floating downstream in a river.   In the atmosphere, winds are not exactly geostrophic, λ can be treated as constant only for the smaller-scale disturbances, and vorticity is always being created or dissipated by friction.   For a first approximation we may, however, ignore these complications; for example, surface friction would require at least a week to bring normal atmospheric motion to rest in the absence of an external source of energy (the sun), and thus is unlikely to affect seriously the validity of conclusions derived from the simplified equations if the forecast is restricted to periods not exceeding, say, 48 hours. The conservation of absolute vorticity provides a starting point for what is now called *numerical forecasting*, the deduction of pressure fields a short time ahead of a given initial distribution by purely mathematical methods.

Long Waves in the Atmosphere

Before proceeding further, it is necessary to examine the physical realities of the problem and to inquire if the equations are likely to yield the kind of information that is useful in forecasting weather. Disturbances in the atmosphere have horizontal scales of length varying from a fraction of a centimetre to thousands of kilometres. At the lower end there are the ripples of pressure that constitute sound waves, then micrometeorological entities such as cloud elements and the eddies that make up the turbulence (or gustiness) of the wind near the surface, and so on, up to the 'sub-synoptic' systems that give rise to local showers and thunderstorms. In general, none of these disturbances can be 'seen' on the meteorologist's synoptic chart, which can reveal only features ranging from small depressions (including tropical cyclones) and anticyclones to the huge semi-permanent continental anticyclones and the long wave-like oscillations in the upper westerlies. More precisely, the features to which the operational forecaster can pay attention are those with horizontal scales of the order of 10^5 metres (50 miles) at least. Smaller disturbances of the pressure field are below the resolving power of the network, because of the considerable interpolation required to construct a synoptic chart from relatively widely spaced observations.

Such large-scale disturbances have one feature in common—they all travel relatively slowly over the surface of the earth. The mean speed of depressions across the British Isles is between 10 and 15 metres a second (20–30 miles an hour). (This refers, of course, to the movement of some recognizable feature of a disturbance, such as its centre, and not to the winds of the system.) In addition, such disturbances usually have something of the character of external waves, in that the shape of the streamlines is much the same at all levels, and the variation of wind with height inside the system does not change greatly from one vertical to another. Fo these reasons, two-dimensional models furnish a tolerable approximation to weather-making disturbances, provided that it is possible to restrict attention to slowly moving systems, that is, to those whose speed is much less than that of other characteristic atmospheric disturbances, such as elastic (sound) and gravitational waves. The translational speed of a synoptic-scale disturbance is not more than about one-thirtieth of the speed of such waves.

The question must now be asked if the existence of such large slowly moving disturbances in the atmosphere can be deduced from the simplified equations of motion and also, if there is any 'preferred pattern' of such disturbances. (In other words, we inquire if the idealization has preserved sufficient features of the real system to be of use in practical problems.) Evidence that long waves exist in such a hypothetical atmosphere was produced in 1939 by the Swedish meteorologist C-G. Rossby. He argued that in purely horizontal motion, the change of relative vorticity with latitude must give rise to characteristic waves in steady or nearly steady motion, and he was able to demonstrate that such oscillations have wavelengths of the required order of magnitude. The proof is very simple. Consider a uniform westerly current of steady velocity U with small unsteady perturbations u' and v', so that if u, v, and ζ are the components of velocity and vorticity, respectively, of the motion as a whole, then

$$u = U + u', \quad v = v', \quad \zeta = \zeta' = \frac{\partial v'}{\partial x} - \frac{\partial u'}{\partial y}$$

The equation for the conservation of absolute vorticity becomes

$$\frac{\partial \zeta}{\partial t} + u\frac{\partial \zeta}{\partial x} + v\frac{\partial \zeta}{\partial y} = -\frac{D\lambda}{Dt} = -v'\frac{\partial \lambda}{\partial y} = -\beta v'$$

where $\beta = \partial \lambda / \partial y = (2/R)\omega \cos \phi$ (R = radius of the earth) prescribes the rate of change of the Coriolis parameter with latitude. If second-order terms of the type $u'\partial \zeta / \partial x$, etc., are disregarded, the equation becomes

$$\frac{\partial \zeta}{\partial t} + U\frac{\partial \zeta}{\partial x} = -\beta v'$$

We are interested primarily in disturbances travelling without change of shape at a constant speed (c), for which we can write $\partial\zeta/\partial t = -c\partial\zeta/\partial x$, and such that the perturbations vary only in the direction of the main stream (i.e., are independent of y). Thus $\zeta = \partial v'/\partial x$ and the equation takes the simpler form

$$(U - c) \frac{\partial^2 v'}{\partial x^2} = -\beta v'$$

This equation has wave-form solutions and if we put $v' = \sin \frac{2\pi}{L} (x - ct)$ it follows that

$$c = U - \beta\left(\frac{L}{2\pi}\right)^2 \qquad \ldots(5)$$

where L is the wavelength of the perturbation (see p. 93). To get an idea of the magnitude of L, consider the case of stationary waves whose length L_s is found by putting $c = 0$ in equation (5). This gives

$$L_s = 2\pi\sqrt{(U/\beta)}$$

In latitude 45°, L_s lies between 3000 and 7000 kilometres (2000 to 4500 miles) when U lies between 4 and 20 metres a second (about 8 to 40 miles an hour). This range of wavelength agrees in magnitude with the horizontal scale of the large disturbances observed in the westerlies, a result which shows that the simplified equations do not exclude disturbances of the type encountered in weather forecasting.

But this is not enough. The general equations of motion also admit the possibility of other oscillations, in particular gravity waves, resembling undulations of a free surface, which in the atmosphere usually travel at speeds not very different from that of sound waves (about 350 metres a second or 700 miles an hour). Waves of this type were generated, for example, by the lifting of the atmosphere in the great Krakatoa volcanic explosion of 1883, and were then of sufficiently large amplitude to be recorded by barographs all over the world. In general, however, gravity waves have amplitudes at least an order of magnitude below the pressure changes that accompany the movements of depressions and anticyclones and may be ignored in forecasting weather, but in the mathematical analysis they are an undesirable complication and may seriously impair the results from the point of view of the meteorologist. Fortunately, in 1948 the American mathematician J. G. Charney proved that the principle of the geostrophic balance, when applied to the vorticity equation, acts as a filtering approximation which excludes the possibility of fast-moving gravity waves appearing in the solution. Sound waves (which also do not affect weather, despite the popular belief that the noise of gunfire causes rain) can be eliminated by the assumption of incompressibility. Finally, when finite differences are used for the solution of the equations of motion by numerical analysis, it is necessary to take other precautions to ensure trustworthy results, but such technical details cannot be discussed here.

Numerical Forecasting

In the simplest type of mathematical forecasting, equation (3), which expresses the conservation of absolute vorticity, is combined with the geostrophic approximation in the form (4) to provide a single partial differential equation for $\partial h/\partial t$, the instantaneous rate of change of contour-height. This equation can be solved numerically, given the initial distribution of h and certain boundary conditions which are discussed below. Suppose we know h over a prescribed area from observations made at some zero of time t_0 and have solved the equation for $\partial h/\partial t$. We can then derive reliable values for h at a subsequent time $t_0 + \Delta t$, where Δt is small (say 1 hour), by straightforward extrapolation, e.g., by the formula

$$h(t_0 + \Delta t) \simeq h(t_0) + \Delta t \left(\frac{\partial h}{\partial t}\right)_{t=t_0}$$

With these new values of h the differential equation is solved again, giving values of $\partial h/\partial t$ at time $t_0 + \Delta t$, and the extrapolation process repeated to give the values of $h(t_0 + 2\Delta t)$, . . ., and so on, until enough (real) time has been covered to give a forecast of h of practical value. From the final set of values of h the meteorologist draws a map of the forecast pressure distribution, called a *prebaratic*.

The details of the calculation are as follows. The differential equation for $\partial h/\partial t$ is easily derived from equations (3) and (4) in the form

$$\nabla^2\!\left(\frac{\partial h}{\partial t}\right) = \mathcal{J}(\eta_g, h) \qquad \ldots (6)$$

where η_g is given by equation (4) and $\mathcal{J}$ is a function called the Jacobian, defined by

$$\mathcal{J}(\eta_g, h) = \frac{\partial \eta_g}{\partial x}\frac{\partial h}{\partial y} - \frac{\partial \eta_g}{\partial y}\frac{\partial h}{\partial x}$$

Equation (6) is of a type well known in mathematical physics as Poisson's equation. To solve this equation the Jacobian has to be determined numerically at discrete points in the selected area. Thus we need to know h everywhere at the zero of time and $\partial h/\partial t$ on the boundaries of the area at all times.

For practical applications, the meteorologist confines his attention to the behaviour of the atmosphere within a large rectangle (e.g. that lying between latitudes 40° and 70° N., and longitudes 40° W. and 40° E.). For the purposes of calculation this area is covered by a grid of mesh size about 300 km. The initial step is to construct, from the observations, the contours of say, the 500-mb surface inside this area and then, by interpolation, to deduce the values of h at the grid points. To evaluate the Jacobian, $\nabla^2(h)$ is found at any grid point (0), approximately, by a finite-difference formula such as

$$\nabla^2(h_0) \simeq \frac{h_1 + h_2 + h_3 + h_4 - 4h_0}{d^2}$$

where h_1, h_2, ... etc., are the values of h at the surrounding grid points and d is the grid spacing. In this way the right-hand side of equation (6) is evaluated at all the grid points at the beginning of each calculation. For the boundary condition we may assume, in theory, any values of $\partial h/\partial t$, and in many of the earlier trials of the method the convenient but unrealistic condition $\partial h/\partial t = 0$ (no change of contour-height on the boundaries during the period of the forecast) was selected. This meant that the solution could be compared with actuality only in the central portions of the area.

The Poisson equation can then be solved at all the grid points using finite-difference methods with an electronic digital computer. The step-by-step process of solution described above is then followed to produce the forecast distribution of h over the rectangle at the end of the prescribed period. A modern electronic computer can tabulate the solution for a 24-hour forecast, involving about 250 grid points, in about an hour, or even less. A human being, working continuously with a desk calculating-machine, would take many months to complete the same task.

The model described above is barotropic; it makes no reference to temperature and obviously represents a very severe idealization of processes in the real atmosphere. Physically, the process of solution amounts to no more than a deduction of changes in the pressure field that result from the simple conservative advection (horizontal motion) of vorticity. The weather forecast, which must contain references to temperature, winds, precipitation and clouds, has to be deduced from the computed pressure distribution by the forecaster, using time-honoured methods. This is often the most uncertain part of the whole process. Yet investigations have shown that even this simple two-dimensional non-divergent model can predict most of the changes that are observed on the 500-mb surface, and a barotropic model has been used, with a fair degree of success, in operational forecasts in Sweden.

The researches undertaken by the Meteorological Office, and by the U.S. Weather Bureau, involve more elaborate models which introduce a certain amount of baroclinicity. In a barotropic model, changes in the pressure field result from the advection of vorticity by the mean flow and the variation of the Coriolis parameter λ with latitude. Baroclinic changes are primarily associated with the advection of the vorticity of the thermal winds. In the Meteorological Office investigations two simultaneous differential equations are used, relating to changes in contour-height and in thickness, respectively, but the analysis is far too lengthy and involved to be given here.

Fig. 54 shows the result of a computation made by Meteorological Office staff for a situation which had given difficulties when treated by the ordinary methods of operational forecasting.* This is a good example of a 24-hour 'machine-

* This example is taken from a paper by I. J. W. Pothecary and F. H. Bushby, *Meteorological Magazine*, 85, 133–142 (1956).

made' forecast of conditions at about 18,000 feet over north-west Europe.   Comparison between the actual and computed charts shows that the major features of the pressure distribution

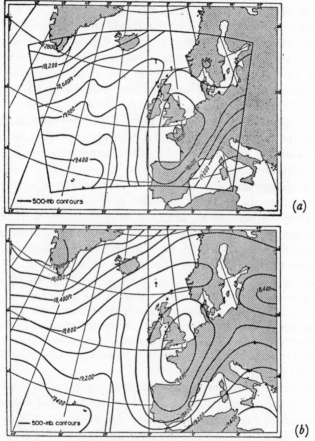

FIG. 54.   Numerical Forecasting.   Computed (a) and actual (b) contour heights for 21 August 1954.

have been forecast by the mathematical process with substantial accuracy, but there are differences of detail.   Such discrepancies are bound to occur, partly because the model does not include all the physical features of the real atmosphere, and partly because of the paucity of the data from which the initial distribution was constructed.

The reader may ask why such calculations are not extended

to longer periods, to produce forecasts for, say, weeks or even months ahead. There is no mathematical difficulty in doing this, but the results would be of no practical value in forecasting weather. Existing models are far too simple for long-range forecasts and, in addition, the difficult question of stability must be considered. A system is said to be stable if no perturbation of the initial state is capable of growing indefinitely—in other words, a stable system possesses inherent qualities which can damp out all disturbances in time. In most stable systems the disturbances are obliterated long before they affect the main motion significantly, and such systems are generally amenable to mathematical analysis. They are *determinate* in the sense that the influence of the initial conditions persists for all time. In unstable systems, on the other hand, disturbances which initially are below the threshold of observation ultimately may take charge and change the character of the whole system as time proceeds. This is so in the atmosphere.

These observations indicate why it is unlikely that any mathematical method can be evolved which will allow accurate predictions of weather to be made for long periods. To quote a present-day worker on atmospheric dynamics (E. T. Eady):

Certain apparently sensible questions, such as the question of weather conditions at a given time in the comparatively distant future, say several days ahead, are *in principle* unanswerable and the most we can hope to do is to determine the relative *probabilities* of different outcomes. The full significance of our theoretical problems becomes apparent only when it is clear what *kind* of question we should attempt to answer.

A hundred years ago it would have been an unthinkable heresy for a scientist to admit that any physical system could defy exact mathematical analysis for all time; to-day we are more disposed to accept the idea that complexity may conceal indeterminacy.

APPENDIX

I. *Arithmetical definition of a limit.* A function $f(x)$ is said to tend to a limit A as x tends to a if we can make the difference between $f(x)$ and A, irrespective of sign, 'as small as we please' by taking x near to a. In strict arithmetical language this is expressed as follows: $f(x)$ tends to the limit A as x tends to a if, given a number ϵ, we can always find an associated number δ so that $|f(x) - A|$ is less than ϵ when $|x - a|$ is less than δ. (The vertical lines indicate that the sign of the differences is ignored.)

The point of the definition is that we imagine an opponent who challenges the statement that the difference between $f(x)$ and A can be made 'as small as we please' by a suitable choice of x. He must name a small number (ϵ) and we must then show that we can always find a value of x sufficiently close to a to make $|f(x) - A|$ less than his ϵ. See G. H. Hardy, *Pure Mathematics* (Cambridge, 1946), for a complete discussion of limits.

II. *Arithmetical definition of continuity.* A function $f(x)$ is said to be continuous at a point $x = a$ if, given a number ϵ, we can always find an associated number δ so that $|f(x) - f(a)|$ is less than ϵ whenever $|x - a|$ is less than δ. This definition implies that

(i) $f(x)$ has a definite value $f(a)$ at $x = a$
(ii) $f(x)$ is defined for all points near $x = a$
(iii) as we approach $x = a$ on either side (i.e. by values of x greater or less than a) $f(x)$ tends to the same limit $f(a)$.

The fundamental property of a continuous function is that which is expressed in everyday language by the statement that there are no 'breaks' in the graph. In precise language: if $y = f(x)$ is continuous at all points between $x = a$ and $x = b$, then as x changes from a to b, y takes at least once every value between $f(a)$ and $f(b)$.

III. *Derivation of the equation of conduction of heat.* As an example of the derivation of a partial differential equation of physics from first principles, consider the flow of heat in an isotropic solid of conductivity k, specific heat c and density ρ. The fundamental law is that the component of the flow of heat (q) in the x-direction across a surface of area A is given by

$$q = - kA\frac{\partial T}{\partial x}$$

where T is temperature. Choose a system of perpendicular axes Ox, Oy, Oz and construct an infinitesimal rectangular box, of sides

δx, δy, δz, around a point (x, y, z). If T is the temperature at (x, y, z), the temperatures on the faces 1 and 2 of the box are, to a first approximation,

$$T - \frac{1}{2}\frac{\partial T}{\partial x}\delta x \quad \text{and} \quad T + \frac{1}{2}\frac{\partial T}{\partial x}\delta x$$

respectively. Hence the flow of heat in the positive x direction through the two faces in unit time is

$$-k\delta y\delta z\frac{\partial}{\partial x}\left(T - \frac{1}{2}\frac{\partial T}{\partial x}\delta x\right) \quad \text{and} \quad -k\delta y\delta z\frac{\partial}{\partial x}\left(T + \frac{1}{2}\frac{\partial T}{\partial x}\delta x\right)$$

respectively, using the fundamental law with $A = \delta y\delta z$.

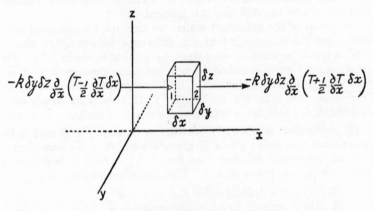

The difference between these two flows is

$$k\frac{\partial^2 T}{\partial x^2}\delta x\delta y\delta z$$

and this must be the heat gained by the infinitesimal element because of the x-component of heat flow. Similarly, the other two faces give

$$k\frac{\partial^2 T}{\partial y^2}\delta x\delta y\delta z \quad \text{and} \quad k\frac{\partial^2 T}{\partial z^2}\delta x\delta y\delta z$$

respectively. Thus the total gain (or loss) of heat by the element in unit time is

$$k\left(\frac{\partial^2 T}{\partial x^2} + \frac{\partial^2 T}{\partial y^2} + \frac{\partial^2 T}{\partial z^2}\right)\delta x\delta y\delta z$$

On the other hand, the gain (or loss) of heat by the element is expressed by

$$\frac{\partial}{\partial t}(c\rho T\delta x\delta y\delta z) = c\rho\frac{\partial T}{\partial t}\delta x\delta y\delta z$$

since $c\rho T$ is, by definition, the heat content of a body of unit volume at uniform temperature T. Equating these two expressions we get

$$c\rho \frac{\partial T}{\partial t} \delta x \delta y \delta z = k \left(\frac{\partial^2 T}{\partial x^2} + \frac{\partial^2 T}{\partial y^2} + \frac{\partial^2 T}{\partial z^2} \right) \delta x \delta y \delta z$$

or

$$\frac{\partial T}{\partial t} = \frac{k}{c\rho} \left(\frac{\partial^2 T}{\partial x^2} + \frac{\partial^2 T}{\partial y^2} + \frac{\partial^2 T}{\partial z^2} \right)$$

This is the equation of heat conduction. The quantity $k/c\rho$ is usually denoted by the Greek letter κ and is called the *thermometric conductivity*.

IV. Bashforth's account of his investigations into air resistance is given in a pamphlet entitled *An Historical Sketch of the Experimental Determination of the Resistance of the Air to the Motion of Projectiles* (Cambridge, 1903). In addition to the purely scientific narrative, the pamphlet contains a lively account of Bashforth's struggles to overcome official prejudice when he began his work in 1864. He says, 'The President and Vice-President of the Ordnance Select Committee were opposed to everything I proposed to do. They professed to know all about ballistics. . . . They had made all necessary experiments and were prepared to furnish me with any amount of results from their own stores! It now became evident that if anything was to be done I must do it myself.'

It is clear, however, that despite the antagonism of the Select Committee, Bashforth received considerable support from the War Office and from the scientific world in general, so much so that he was encouraged to allow his name to go forward for election to the Royal Society. His failure to receive this recognition was a sore blow and, characteristically, he ascribes his rejection to action by his 'enemies'. He quotes a letter from Sir George Airy dated 6th April, 1870, which seems to support this view. Airy says, 'In the list of Candidates for the Royal Society which has just reached me I see your name. Your certificate has never been before me and when I took my round in the Society's Meeting Room expressly to look at certificates, I am most confident it was not there. Had I seen it I should assuredly have given myself the pleasure of signing it.' Bashforth's comment on this odd episode is, 'I troubled no more about the matter. I left my enemies to enjoy their triumph—such as it was.'

Bashforth resigned his professorship in 1874. He gives his reasons for taking this step in the following words: 'March 12, 1874, my particular attention was officially directed to Articles 3 and 4 of a new Royal Warrant of February 12, 1874. As this document indicated a small reduction in my small stipend, I at once retired. . . . Considering the *success* of my work, the *labour* it had entailed during the preceeding *ten* years and the *advantage* to the government, a stipend three or four times what I received would not have been too much in return for the services rendered by me to the government.'

Modern ballistic research is extremely expensive. Bashforth's experiments were criticized on the grounds that they were 'the most costly ever known'. He points out, in reply, that he built up his tables on the results from some 500 rounds, at a total cost in the neighbourhood of £1000 saying, quite justly, that this was a small outlay for such valuable information. To-day, an expenditure of £1000 is regarded as very low for any fundamental work in armament science.

A photograph of Bashforth in the possession of the Royal Military College of Science shows him as a heavily whiskered clergyman of benevolent appearance, the typical country parson of his day. After so many years, it is impossible to judge fairly his complaints against officialdom, but there can be no doubt that he stood head and shoulders above his contemporary ballisticians, both as regards mathematical ability and experimental skill.

V. *Siacci's derivation of the ballistic equations.* Equations (7) to (10) give

$$\frac{d}{dt}(V \cos \theta) = -r \cos \theta \qquad \ldots(A)$$

$$\frac{d}{dt}(V \sin \theta) = -r \sin \theta - g \qquad \ldots(B)$$

Elimination of dV/dt and r gives

$$V\frac{d\theta}{dt} = -g \cos \theta$$

which is equation (14). Also, from Eq. (A) above

$$\frac{d}{dt}(V \cos \theta) = \frac{d}{d\theta}(V \cos \theta)\frac{d\theta}{dt}$$

$$= \frac{d}{d\theta}(V \cos \theta) - g \frac{\cos \theta}{V} = r \cos \theta$$

or

$$\frac{d}{d\theta}(V \cos \theta) = \frac{rV}{g} = \frac{\kappa\rho V^3}{C_0 g} \qquad \ldots(C)$$

i.e. Eq. (11). For small values of θ, $\cos \theta$ is nearly unity and equation (C) becomes

$$\frac{dV}{d\theta} = \frac{\kappa\rho V^3}{C_0 g}$$

which is the hodograph equation (IIa). The remaining equations are obvious.

VI. *The Newton–Raphson iterative process for finding the numerical value of a root of an equation.* Let the equation be denoted by $f(x) = 0$, and let $x = x_0$ be an approximation to the desired root. In general,

$$x_0 - \frac{f(x_0)}{f'(x_0)}$$

is a better approximation to the true value of the root, and the process may be repeated any number of times. It is necessary that x_0 should be a fairly close approximation for results of real value.

In the practical application it is often more convenient to express the rule as follows: if the required root is known to lie between $x = a$ and $x = a + h$, where h is small, a good approximation to the true value is

$$a + \frac{hf(a)}{f(a) - f(a + h)}$$

If this is not adequate, the process is applied again using the new approximation as a starting point. The value of this form lies in the fact that usually it is easy to locate a root approximately by finding two values of x for which $f(x)$ has opposite signs (a continuous function changes sign when passing through zero). The first step is to find two neighbouring values of x which make $f(x)$ a small positive and a small negative number, respectively.

As an example consider the equation

$$f(x) \equiv x^3 + x - 3 = 0$$

which has a real root not very different from unity. To find the value of this root correct to, say, two places of decimals, we note that $f(1) = -1$ and $f(1\frac{1}{2}) = +\frac{15}{8}$. Hence the root must lie between 1 and 1·5. Take $a = 1$, $h = \frac{1}{2}$ and apply the rule. The approximate value is

$$1 + \frac{\frac{1}{2}(-1)}{-1 - \frac{15}{8}} = 1\frac{4}{23} = 1 \cdot 18 \text{ (to two places)}$$

Actually, this is quite a good approximation to the true value (1·21), but the process should be repeated (taking, say, $a = 1 \cdot 18$ and $h = 0 \cdot 07$) and so on, until further applications of the rule do not change the value of the second place of decimals.

VII. *Calculation of the aerodynamic force on a cylinder immersed in a steady uniform stream of incompressible inviscid fluid.* To illustrate the method of finding the aerodynamic force described on p. 146 consider the above example. The various steps are as follows:

(i) *The complex potential.* This is

$$w(z) = V\left(z + \frac{a^2}{z}\right)$$

where a is real. This function is holomorphic at all points except $z = 0$; V is the velocity 'at infinity'.

(ii) *The streamfunction.* Put $z = x + iy$. The complex potential becomes

$$w = \phi + i\psi = V\left(x + iy + \frac{a^2}{x + iy}\right)$$

$$= V\left\{x + iy + \frac{a^2(x - iy)}{x^2 + y^2}\right\}$$

and the imaginary part is

$$\psi(x, y) = V\left(y - \frac{a^2y}{x^2 + y^2}\right)$$

If we exclude the origin, ψ is the streamfunction of a possible irrotational motion. As $x, y \to \infty$, $\psi \to Vy$, so that in parts of the plane remote from the origin the motion is indistinguishable from that of a uniform stream of velocity V parallel to the x-axis (i.e. the streamlines are the lines $y = $ constant). The streamline $\psi = 0$ gives

$$y = 0 \quad \text{if} \quad x \neq 0$$

and

$$x^2 + y^2 = a^2$$

and thus consists of the axis of x and the circle $x^2 + y^2 = a^2$, centre $(0, 0)$, and radius a. Outside the circle the streamlines are curves intermediate between the straight lines of the uniform stream, the circle and the x-axis. This is the pattern of Fig. 36. We are not concerned with points inside the circle.

(iii) *The velocity field.* It is convenient here to change to polar coordinates. Put $x = r \cos \theta, y = r \sin \theta$. Then

$$\psi(r, \theta) = V\left(r - \frac{a^2}{r}\right) \sin \theta$$

Instead of velocity components parallel to the axes of x and y we have radial (u_r) and tangential (u_t) components. These are

$$u_r = \frac{1}{r} \frac{\partial \psi}{\partial \theta} = V\left(1 - \frac{a^2}{r^2}\right) \cos \theta \qquad \text{...}(A)$$

$$u_t = -\frac{\partial \psi}{\partial r} = V\left(1 + \frac{a^2}{r^2}\right) \sin \theta \qquad \text{...}(B)$$

The circle may now be interpreted as the cross section of a cylinder of infinite length placed across stream. On the circle $r = a$ and thus, from eqs. A and B

$$u_r = 0; \quad u_t = 2V \sin \theta$$

The radial velocity is necessarily zero because fluid cannot cross a streamline. The tangential velocity is zero at the points where $\theta = 0$ and 2π and reaches a maximum value $2V$ at points where $\theta = \frac{1}{2}\pi$ and $\frac{3}{2}\pi$.

(iv) *The pressure field.* Apply Bernoulli's theorem to the streamline $\psi = 0$ which originates in a region of the field where $u_r = V$ and $p = p_0$, say. Hence, if p is the pressure at any point (a, θ) on the circle,

$$p + \tfrac{1}{2}\rho(2V \sin \theta)^2 = p_0 + \tfrac{1}{2}\rho V^2$$

(v) *The aerodynamic force.* Consider an element of arc of the circle, length $ad\theta$. If X and Y are the components (parallel to the x and y axes, respectively) of the resultant thrust caused by the pressure distribution,

$$dX = -\, pad\theta \cos \theta; \quad dY = -\, pad\theta \sin \theta$$

Over the whole circle

$$X = -\int_0^{2\pi} pa \cos \theta d\theta; \quad Y = -\int_0^{2\pi} pa \sin \theta d\theta$$

On substitution of the value of p from Bernoulli's theorem it follows that

$$X = -\int_0^{2\pi} a \cos \theta(p_0 + \tfrac{1}{2}\rho V^2 - 2\rho V^2 \sin^2 \theta)d\theta$$

$$Y = -\int_0^{2\pi} a \sin \theta(p_0 + \tfrac{1}{2}\rho V^2 - 2\rho V^2 \sin^2 \theta d\theta)d\theta$$

and by integration it follows that $X = Y = 0$ because

$$\int_0^{2\pi} \sin \theta d\theta, \int_0^{2\pi} \cos \theta d\theta, \int_0^{2\pi} \cos \theta \sin^2 \theta d\theta \text{ and } \int_0^{2\pi} \sin^3 \theta d\theta$$

are all zero.

In the example of the Magnus effect the complex potential is

$$w(z) = V\left(z + \frac{a^2}{z}\right) + \frac{iK}{2\pi} \log\left(\frac{z}{a}\right)$$

where K is the circulation.

The radial component of velocity is the same as in the previous example, but the tangential component has the additional term $K/2\pi r$. In the subsequent analysis $X = 0$ as before (no drag) but

$$Y = \frac{\rho K V}{\pi}\int_0^{2\pi} \sin^2 \theta d\theta = \rho K V$$

which is an example of the Kutta-Joukowski theorem.

VIII. *Joukowski transformations.* The Joukowski mapping function is

$$w = z + \frac{b^2}{z}$$

(and is thus similar to the complex potential for streaming flow past a circle). The circle to be transformed always includes the origin. The points $z = \pm b$ are called the singular points of the transformation

(i) *Circle centred at* $z = 0$. In general, this circle is transformed into an ellipse. If the circle passes through the singular points it transforms into a part of the real axis.

(ii) *Circle centred on the imaginary axis.* If the circle passes through the singular points, it transforms into a circular arc with the part of the real axis between $\pm 2b$ as chord. The transformed curve is to be interpreted as the profile of a thin curved aerofoil.

(iii) *Circle centred on the real axis.* The case of greatest practical interest is when the radius of the original circle is slightly greater than b and the circle passes through one singular point, the other lying inside the circle. The transformed curve is symmetrical and pear-shaped, with a cusp. The axis of symmetry (called the *skeleton*) is the part of the real axis between $\pm 2b$, and the position of the point of maximum thickness is b from the thick leading edge. Shapes of this type are used as rudders or fins.

(iv) *Circle centred at a point not on either axis.* In this case the Joukowski aerofoil profile is obtained. This may be regarded, somewhat crudely, as the pear-shaped section (iii) with the skeleton bent into a curved line. The sharply tapered trailing edge is formed by making the original circle pass through one of the singular points. For a practicable profile the centre of the circle needs to be placed fairly close to the imaginary axis.

In the *Kármán–Trefftz* transformations the upper and lower surfaces of the aerofoil form a small (non-zero) angle at the trailing edge. In the *Carafoli transformations* the sharp point at the tail is replaced by a rounded termination. The *Mises type of transformation* gives profiles with pointed tails with skeletons having points of inflexion. Details of these transformations are to be found in *Aerodynamic Theory*, ed. Durand, Division E (1934), contributed by Kármán and Burgers.

IX. *Fitting a line by the method of least squares.* Suppose we have n

points indicated by $(x_1, y_1), (x_2, y_2), \ldots, (x_n, y_n)$, to which it is desired to fit the straight line

$$y = ax + b$$

by the method of least squares. The numbers a and b are found by solving the simultaneous equations

$$\Sigma y_i = na + b\Sigma x_i$$
$$\Sigma x_i y_i = a\Sigma x_i + b\Sigma x_i{}^2$$

Thus the computer has first to work out

$$\Sigma x_i = x_1 + x_2 + \ldots + x_n$$
$$\Sigma y_i = y_1 + y_2 + \ldots + y_n$$
$$\Sigma x_i y_i = x_1 y_1 + x_2 y_2 + \ldots + x_n y_n$$
$$\Sigma x_i{}^2 = x_1{}^2 + x_2{}^2 + \ldots + x_n{}^2$$

X. *Calculation of probabilities in coin-tossing trials.* Coin-tossing illustrates the use of the binomial distribution in the theory of probability. This says that if p is the chance of success, and q the chance of failure, in a single trial, the chances of $0, 1, 2, \ldots, n$ successes in n independent trials are given by the terms of the binomial expansion $(p + q)^n$. When tossing a coin the chance of getting (or not getting) a head or a tail is clearly $\frac{1}{2}$, so that here $p = q = \frac{1}{2}$. The chances of getting $0, 1, 2, \ldots, 10$ heads (or tails) in 10 tosses are:

$$(\tfrac{1}{2} + \tfrac{1}{2})^{10} \simeq 0.001 + 0.010 + 0.044 + 0.117 + 0.205 + 0.246 +$$
$$0.205 + 0.117 + 0.044 + 0.010 + 0.001$$

(For example, the chance of 4 tails in 10 tosses is 0.205, or about 1 in 5.)

In tests to decide whether a coin is biased, such as those described on p. 189, we are interested only in the degree of asymmetry of the result. Here we must consider the probability of a result which is at least as uneven as that obtained, e.g., in the example quoted (7 heads and 3 tails) we must sum the chances of getting 7, 8, 9, 10 heads and 7, 8, 9, 10 tails. Hence the relevant probability is

$$2(0.001 + 0.010 + 0.044 + 0.117) = 0.344$$

or about 34 per cent.

INDEX

A CATALOGUE OF SELECTED DOVER BOOKS
IN ALL FIELDS OF INTEREST

A CATALOGUE OF SELECTED DOVER
BOOKS IN ALL FIELDS OF INTEREST

CELESTIAL OBJECTS FOR COMMON TELESCOPES, T. W. Webb. The most used book in amateur astronomy: inestimable aid for locating and identifying nearly 4,000 celestial objects. Edited, updated by Margaret W. Mayall. 77 illustrations. Total of 645pp. 5⅜ x 8½.
20917-2, 20918-0 Pa., Two-vol. set $9.00

HISTORICAL STUDIES IN THE LANGUAGE OF CHEMISTRY, M. P. Crosland. The important part language has played in the development of chemistry from the symbolism of alchemy to the adoption of systematic nomenclature in 1892. ". . . wholeheartedly recommended,"—Science. 15 illustrations. 416pp. of text. 5⅝ x 8¼.   63702-6 Pa. $6.00

BURNHAM'S CELESTIAL HANDBOOK, Robert Burnham, Jr. Thorough, readable guide to the stars beyond our solar system. Exhaustive treatment, fully illustrated. Breakdown is alphabetical by constellation: Andromeda to Cetus in Vol. 1; Chamaeleon to Orion in Vol. 2; and Pavo to Vulpecula in Vol. 3. Hundreds of illustrations. Total of about 2000pp. 6⅛ x 9¼.
23567-X, 23568-8, 23673-0 Pa., Three-vol. set $27.85

THEORY OF WING SECTIONS: INCLUDING A SUMMARY OF AIR-FOIL DATA, Ira H. Abbott and A. E. von Doenhoff. Concise compilation of subatomic aerodynamic characteristics of modern NASA wing sections, plus description of theory. 350pp. of tables. 693pp. 5⅜ x 8½.
60586-8 Pa. $8.50

DE RE METALLICA, Georgius Agricola. Translated by Herbert C. Hoover and Lou H. Hoover. The famous Hoover translation of greatest treatise on technological chemistry, engineering, geology, mining of early modern times (1556). All 289 original woodcuts. 638pp. 6¾ x 11.
60006-8 Clothbd. $17.95

THE ORIGIN OF CONTINENTS AND OCEANS, Alfred Wegener. One of the most influential, most controversial books in science, the classic statement for continental drift. Full 1966 translation of Wegener's final (1929) version. 64 illustrations. 246pp. 5⅜ x 8½.   61708-4 Pa. $4.50

THE PRINCIPLES OF PSYCHOLOGY, William James. Famous long course complete, unabridged. Stream of thought, time perception, memory, experimental methods; great work decades ahead of its time. Still valid, useful; read in many classes. 94 figures. Total of 1391pp. 5⅜ x 8½.
20381-6, 20382-4 Pa., Two-vol. set $13.00

TONE POEMS, SERIES II: TILL EULENSPIEGELS LUSTIGE STREICHE, ALSO SPRACH ZARATHUSTRA, AND EIN HELDEN-LEBEN, Richard Strauss. Three important orchestral works, including very popular *Till Eulenspiegel's Marry Pranks*, reproduced in full score from original editions. Study score. 315pp. 9⅜ x 12¼. (Available in U.S. only)
23755-9 Pa. $8.95

TONE POEMS, SERIES I: DON JUAN, TOD UND VERKLARUNG AND DON QUIXOTE, Richard Strauss. Three of the most often performed and recorded works in entire orchestral repertoire, reproduced in full score from original editions. Study score. 286pp. 9⅜ x 12¼. (Available in U.S. only)
23754-0 Pa. $7.50

11 LATE STRING QUARTETS, Franz Joseph Haydn. The form which Haydn defined and "brought to perfection." (*Grove's*). 11 string quartets in complete score, his last and his best. The first in a projected series of the complete Haydn string quartets. Reliable modern Eulenberg edition, otherwise difficult to obtain. 320pp. 8⅜ x 11¼. (Available in U.S. only)
23753-2 Pa. $7.50

FOURTH, FIFTH AND SIXTH SYMPHONIES IN FULL SCORE, Peter Ilyitch Tchaikovsky. Complete orchestral scores of Symphony No. 4 in F Minor, Op. 36; Symphony No. 5 in E Minor, Op. 64; Symphony No. 6 in B Minor, "Pathetique," Op. 74. Bretikopf & Hartel eds. Study score. 480pp. 9⅜ x 12¼.
23861-X Pa. $10.95

THE MARRIAGE OF FIGARO: COMPLETE SCORE, Wolfgang A. Mozart. Finest comic opera ever written. Full score, not to be confused with piano renderings. Peters edition. Study score. 448pp. 9⅜ x 12¼. (Available in U.S. only)
23751-6 Pa. $11.95

"IMAGE" ON THE ART AND EVOLUTION OF THE FILM, edited by Marshall Deutelbaum. Pioneering book brings together for first time 38 groundbreaking articles on early silent films from *Image* and 263 illustrations newly shot from rare prints in the collection of the International Museum of Photography. A landmark work. Index. 256pp. 8¼ x 11.
23777-X Pa. $8.95

AROUND-THE-WORLD COOKY BOOK, Lois Lintner Sumption and Marguerite Lintner Ashbrook. 373 cooky and frosting recipes from 28 countries (America, Austria, China, Russia, Italy, etc.) include Viennese kisses, rice wafers, London strips, lady fingers, hony, sugar spice, maple cookies, etc. Clear instructions. All tested. 38 drawings. 182pp. 5⅜ x 8.
23802-4 Pa. $2.50

THE ART NOUVEAU STYLE, edited by Roberta Waddell. 579 rare photographs, not available elsewhere, of works in jewelry, metalwork, glass, ceramics, textiles, architecture and furniture by 175 artists—Mucha, Seguy, Lalique, Tiffany, Gaudin, Hohlwein, Saarinen, and many others. 288pp. 8⅜ x 11¼.
23515-7 Pa. $6.95

THE PHILOSOPHY OF HISTORY, Georg W. Hegel. Great classic of Western thought develops concept that history is not chance but a rational process, the evolution of freedom. 457pp. 5⅜ x 8½. 20112-0 Pa. $4.50

LANGUAGE, TRUTH AND LOGIC, Alfred J. Ayer. Famous, clear introduction to Vienna, Cambridge schools of Logical Positivism. Role of philosophy, elimination of metaphysics, nature of analysis, etc. 160pp. 5⅜ x 8½. (Available in U.S. only) 20010-8 Pa. $2.00

A PREFACE TO LOGIC, Morris R. Cohen. Great City College teacher in renowned, easily followed exposition of formal logic, probability, values, logic and world order and similar topics; no previous background needed. 209pp. 5⅜ x 8½. 23517-3 Pa. $3.50

REASON AND NATURE, Morris R. Cohen. Brilliant analysis of reason and its multitudinous ramifications by charismatic teacher. Interdisciplinary, synthesizing work widely praised when it first appeared in 1931. Second (1953) edition. Indexes. 496pp. 5⅜ x 8½. 23633-1 Pa. $6.50

AN ESSAY CONCERNING HUMAN UNDERSTANDING, John Locke. The only complete edition of enormously important classic, with authoritative editorial material by A. C. Fraser. Total of 1176pp. 5⅜ x 8½. 20530-4, 20531-2 Pa., Two-vol. set $16.00

HANDBOOK OF MATHEMATICAL FUNCTIONS WITH FORMULAS, GRAPHS, AND MATHEMATICAL TABLES, edited by Milton Abramowitz and Irene A. Stegun. Vast compendium: 29 sets of tables, some to as high as 20 places. 1,046pp. 8 x 10½. 61272-4 Pa. $14.95

MATHEMATICS FOR THE PHYSICAL SCIENCES, Herbert S. Wilf. Highly acclaimed work offers clear presentations of vector spaces and matrices, orthogonal functions, roots of polynomial equations, conformal mapping, calculus of variations, etc. Knowledge of theory of functions of real and complex variables is assumed. Exercises and solutions. Index. 284pp. 5⅝ x 8¼. 63635-6 Pa. $5.00

THE PRINCIPLE OF RELATIVITY, Albert Einstein et al. Eleven most important original papers on special and general theories. Seven by Einstein, two by Lorentz, one each by Minkowski and Weyl. All translated, unabridged. 216pp. 5⅜ x 8½. 60081-5 Pa. $3.50

THERMODYNAMICS, Enrico Fermi. A classic of modern science. Clear, organized treatment of systems, first and second laws, entropy, thermodynamic potentials, gaseous reactions, dilute solutions, entropy constant. No math beyond calculus required. Problems. 160pp. 5⅜ x 8½. 60361-X Pa. $3.00

ELEMENTARY MECHANICS OF FLUIDS, Hunter Rouse. Classic undergraduate text widely considered to be far better than many later books. Ranges from fluid velocity and acceleration to role of compressibility in fluid motion. Numerous examples, questions, problems. 224 illustrations. 376pp. 5⅝ x 8¼. 63699-2 Pa. $5.00

UNCLE SILAS, J. Sheridan LeFanu. Victorian Gothic mystery novel, considered by many best of period, even better than Collins or Dickens. Wonderful psychological terror. Introduction by Frederick Shroyer. 436pp. 5⅜ x 8½.   21715-9 Pa. $6.00

JURGEN, James Branch Cabell. The great erotic fantasy of the 1920's that delighted thousands, shocked thousands more. Full final text, Lane edition with 13 plates by Frank Pape. 346pp. 5⅜ x 8½.
                                                      23507-6 Pa. $4.50

THE CLAVERINGS, Anthony Trollope. Major novel, chronicling aspects of British Victorian society, personalities. Reprint of Cornhill serialization, 16 plates by M. Edwards; first reprint of full text. Introduction by Norman Donaldson. 412pp. 5⅜ x 8½.   23464-9 Pa. $5.00

KEPT IN THE DARK, Anthony Trollope. Unusual short novel about Victorian morality and abnormal psychology by the great English author. Probably the first American publication. Frontispiece by Sir John Millais. 92pp. 6½ x 9¼.   23609-9 Pa. $2.50

RALPH THE HEIR, Anthony Trollope. Forgotten tale of illegitimacy, inheritance. Master novel of Trollope's later years. Victorian country estates, clubs, Parliament, fox hunting, world of fully realized characters. Reprint of 1871 edition. 12 illustrations by F. A. Faser. 434pp. of text. 5⅜ x 8½.   23642-0 Pa. $5.00

YEKL and THE IMPORTED BRIDEGROOM AND OTHER STORIES OF THE NEW YORK GHETTO, Abraham Cahan. Film Hester Street based on Yekl (1896). Novel, other stories among first about Jewish immigrants of N.Y.'s East Side. Highly praised by W. D. Howells—Cahan "a new star of realism." New introduction by Bernard G. Richards. 240pp. 5⅜ x 8½.   22427-9 Pa. $3.50

THE HIGH PLACE, James Branch Cabell. Great fantasy writer's enchanting comedy of disenchantment set in 18th-century France. Considered by some critics to be even better than his famous Jurgen. 10 illustrations and numerous vignettes by noted fantasy artist Frank C. Pape. 320pp. 5⅜ x 8½.   23670-6 Pa. $4.00

ALICE'S ADVENTURES UNDER GROUND, Lewis Carroll. Facsimile of ms. Carroll gave Alice Liddell in 1864. Different in many ways from final Alice. Handlettered, illustrated by Carroll. Introduction by Martin Gardner. 128pp. 5⅜ x 8½.   21482-6 Pa. $2.50

FAVORITE ANDREW LANG FAIRY TALE BOOKS IN MANY COLORS, Andrew Lang. The four Lang favorites in a boxed set—the complete Red, Green, Yellow and Blue Fairy Books. 164 stories; 439 illustrations by Lancelot Speed, Henry Ford and G. P. Jacomb Hood. Total of about 1500pp. 5⅜ x 8½.   23407-X Boxed set, Pa. $15.95

PRINCIPLES OF ORCHESTRATION, Nikolay Rimsky-Korsakov. Great classical orchestrator provides fundamentals of tonal resonance, progression of parts, voice and orchestra, tutti effects, much else in major document. 330pp. of musical excerpts. 489pp. 6½ x 9¼.   21266-1 Pa. $7.50

TRISTAN UND ISOLDE, Richard Wagner. Full orchestral score with complete instrumentation. Do not confuse with piano reduction. Commentary by Felix Mottl, great Wagnerian conductor and scholar. Study score. 655pp. 8⅛ x 11.   22915-7 Pa. $13.95

REQUIEM IN FULL SCORE, Giuseppe Verdi. Immensely popular with choral groups and music lovers. Republication of edition published by C. F. Peters, Leipzig, n. d. German frontmaker in English translation. Glossary. Text in Latin. Study score. 204pp. 9⅜ x 12¼.
23682-X Pa. $6.00

COMPLETE CHAMBER MUSIC FOR STRINGS, Felix Mendelssohn. All of Mendelssohn's chamber music: Octet, 2 Quintets, 6 Quartets, and Four Pieces for String Quartet. (Nothing with piano is included). Complete works edition (1874-7). Study score. 283 pp. 9⅜ x 12¼.
23679-X Pa. $7.50

POPULAR SONGS OF NINETEENTH-CENTURY AMERICA, edited by Richard Jackson. 64 most important songs: "Old Oaken Bucket," "Arkansas Traveler," "Yellow Rose of Texas," etc. Authentic original sheet music, full introduction and commentaries. 290pp. 9 x 12.   23270-0 Pa. $7.95

COLLECTED PIANO WORKS, Scott Joplin. Edited by Vera Brodsky Lawrence. Practically all of Joplin's piano works—rags, two-steps, marches, waltzes, etc., 51 works in all. Extensive introduction by Rudi Blesh. Total of 345pp. 9 x 12.   23106-2 Pa. $14.95

BASIC PRINCIPLES OF CLASSICAL BALLET, Agrippina Vaganova. Great Russian theoretician, teacher explains methods for teaching classical ballet; incorporates best from French, Italian, Russian schools. 118 illustrations. 175pp. 5⅜ x 8½.   22036-2 Pa. $2.50

CHINESE CHARACTERS, L. Wieger. Rich analysis of 2300 characters according to traditional systems into primitives. Historical-semantic analysis to phonetics (Classical Mandarin) and radicals. 820pp. 6⅛ x 9¼.
21321-8 Pa. $10.00

EGYPTIAN LANGUAGE: EASY LESSONS IN EGYPTIAN HIERO-GLYPHICS, E. A. Wallis Budge. Foremost Egyptologist offers Egyptian grammar, explanation of hieroglyphics, many reading texts, dictionary of symbols. 246pp. 5 x 7½. (Available in U.S. only)
21394-3 Clothbd. $7.50

AN ETYMOLOGICAL DICTIONARY OF MODERN ENGLISH, Ernest Weekley. Richest, fullest work, by foremost British lexicographer. Detailed word histories. Inexhaustible. Do not confuse this with *Concise Etymological Dictionary*, which is abridged. Total of 856pp. 6½ x 9¼.
21873-2, 21874-0 Pa., Two-vol. set $12.00

GEOMETRY, RELATIVITY AND THE FOURTH DIMENSION, Rudolf Rucker. Exposition of fourth dimension, means of visualization, concepts of relativity as Flatland characters continue adventures. Popular, easily followed yet accurate, profound. 141 illustrations. 133pp. 5⅜ x 8½.
23400-2 Pa. $2.75

THE ORIGIN OF LIFE, A. I. Oparin. Modern classic in biochemistry, the first rigorous examination of possible evolution of life from nitrocarbon compounds. Non-technical, easily followed. Total of 295pp. 5⅜ x 8½.
60213-3 Pa. $4.00

PLANETS, STARS AND GALAXIES, A. E. Fanning. Comprehensive introductory survey: the sun, solar system, stars, galaxies, universe, cosmology; quasars, radio stars, etc. 24pp. of photographs. 189pp. 5⅜ x 8½. (Available in U.S. only)
21680-2 Pa. $3.75

THE THIRTEEN BOOKS OF EUCLID'S ELEMENTS, translated with introduction and commentary by Sir Thomas L. Heath. Definitive edition. Textual and linguistic notes, mathematical analysis, 2500 years of critical commentary. Do not confuse with abridged school editions. Total of 1414pp. 5⅜ x 8½.
60088-2, 60089-0, 60090-4 Pa., Three-vol. set $18.50

Available at your book dealer or write for free catalogue to Dept. GI, Dover Publications, Inc., 31 East Second Street, Mineola, N.Y. 11501. Dover publishes more than 175 books each year on science, elementary and advanced mathematics, biology, music, art, literary history, social sciences and other areas.